CALIFORNIA
LIZARDS
AND HOW TO FIND THEM

Western Banded Gecko. *Photograph by Spencer Riffle.*

CALIFORNIA LIZARDS

AND HOW TO FIND THEM

EMILY TAYLOR

H
HEYDAY

Berkeley, California

Library of Congress Cataloging-in-Publication Data is available.

Cover Art: Spencer Riffle
Cover and Interior Design: Debbie Berne

Published by Heyday
PO Box 9145, Berkeley, California 94709
(510) 549-3564
heydaybooks.com

Printed in East Peoria, Illinois, by Versa Press, Inc.

10 9 8 7 6 5 4 3 2 1

To Steve
You had me with the lizard tattoo

CONTENTS

◀ A female Common Side-blotched Lizard rests near a mammal burrow where she takes shelter and will soon lay her eggs. *Photograph by Marisa Ishimatsu.*

Common Chuckwalla.
Photograph by Max Roberts.

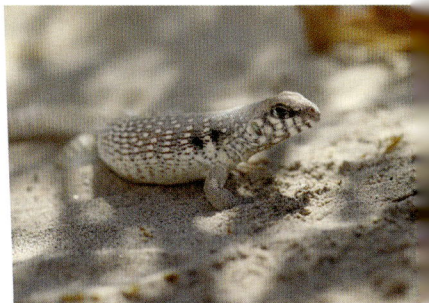

Desert Iguana. *Photograph by Brandon Kong.*

Western Fence Lizard. *Photograph by Zeev Nitzan Ginsburg.*

PREFACE

Do you love lizards? Me, too. And we are not alone. Aside from the small slice of the population that considers *all* tiny animals to be either terrifying or disgusting, most people tend to find lizards endearing, cute, interesting, or at the very least unthreatening. Parents that outright refuse to let their child get a pet snake will buy them a bearded dragon, even though the lizard is *much* harder to properly care for than the snake. The presence of four limbs is undoubtedly reassuring to those suspicious of critters with too few (or too many!) legs.

There are a lot of reasons why people like lizards. We are naturally drawn to wildlife that we consider beautiful and benevolent, even more so if they are helpful to us. Many lizards are colored like bright jewels. Some resemble mini dinosaurs (*dinosaur* means "terrible lizard" in Greek, though dinosaurs were surely neither terrible nor lizards!). Their territorial antics are charming. Collectively, they form an insect-eating army that helps keep pests at bay. Most of them are small. With rare exceptions, they are harmless to people.

Lizards are also familiar. While many people have only ever seen a snake on TV or on the internet, lizards are practically everywhere. You might have grown up surrounded by blue-belly lizards that put on comical territorial displays on the fence posts in your yard that they call home. Perhaps you brought your family T-shirts from your vacation in Cabo San Lucas featuring sunglass-clad Green Iguanas or shot glasses from Hawai'i decked out with geckos (even though neither of these species are native to those areas). Admit it—you laughed heartily at the GEICO gecko's witticisms, despite being annoyed by commercial breaks.

The beauty and familiarity of lizards, especially the ubiquitous blue-bellies, are reflected in the artwork and stories of many

Sean Barefield

It's impossible to deny that this Desert Horned Lizard's face is endearing.

Indigenous groups in California and elsewhere in the Southwest. Lizards represent a spectrum of forces, all good, ranging from prosperity and luck to healing and rebirth. My favorite Native American lizard story explains how the blue-belly lizard got its bright colors. In an article in *Bay Nature* magazine, Greg Sarris recounted this Southern Pomo and Coast Miwok tale:

> *My ancestors always knew the bluebelly was important. He knows the sun better than any of us, after all. According to the old stories, the sun gave Bluebelly a piece of its home–the sky–to wear as a sign of kinship. Sun said to Bluebelly, "With my home on your stomach the people will always know to remember me. When they see you each spring, again sitting atop the rocks, they will know too that I have returned. Your belly will match the sky where once again I'm*

Marisa Ishimatsu

A Western Fence Lizard (a.k.a. "blue-belly") overlooks the San Francisco Bay from its basking perch.

looking down." That's why he's one of the first creatures to come out in the spring, and why he sits where the sun can see him.

I love this story because it resonates with me. Indeed, the blue-belly (a.k.a. Western Fence Lizard) is a harbinger of spring in California, and my first glimpse of them as temperatures rise is always a time for celebration. It's the season to shed sweaters for tank tops, to get those tomato seedlings into the ground, and to prepare for all my favorite scaly creatures to emerge from their long winter sleep.

A story told among multiple Indigenous groups in California credits the blue-belly lizard for the creation of human hands. Some versions of the story say the lizard won a fight against the coyote, who was a creator of sorts, obliging the coyote to give people hands instead of paws. Others say it was a friendlier persuasion, with the lizard advising the coyote to give people hands like a lizard's so that they could use tools. I wrote this book on the seized ancestral homelands of the Salinan people, in whose Migueleño dialect some lizards are known as cwa•kek'á'. I pay homage to the Californians who came before me for their stewardship of these lands, and I thank the cwa•kek'á' for giving me the dexterous fingers with which to type these words about him and his brethren.

While lizards might be familiar, sometimes familiarity breeds oblivion, and harmless and abundant things fade into the background and become part of the landscape. Consider trees for a moment. Many of us are surrounded by ancient behemoths towering over our homes and lining our boulevards, but most of the time we don't even notice them. Similar to this "arboroblivion," many people don't take note of the tiny lizards zigging and zagging around us.

In this book, I hope to conquer your "sauroblivion" by piquing your interest in these fascinating, scaly little beasts. California is home to nearly sixty species of lizards, and many of them are highly watchable. Unable to chat verbally, they perform an assortment of adorable behaviors to communicate with one another, and sometimes even with us, from their perches. A graceful Zebra-tailed Lizard waves its conspicuous black and white tail to inform you that you might as well abandon your attempt to sneak up on it. A non-native, female Jackson's Chameleon rapidly deepens its color from light to dark green when harassed by an amorous male. A similarly hopeful male Granite Spiny Lizard flashes his sapphire-blue chest at a female while performing flamboyant push-ups to warn off an intruding male in the distance.

It's one thing to read about these charming behaviors in a book and quite another to witness them in real life. Maybe you're a Californian who wants to know more about the wildlife around

Chad Lane

A Northern Legless Lizard escapes into the sand.

you. Perhaps you're a visitor excited to identify those lizards scaling the trailside boulders on your treks through California's magnificent regional, state, and national parks. Or maybe this is a gift from a friend or family member who thinks you might appreciate the quirky little critters that frequent most parts of our beautiful state.

This book is not a field guide. While it does contain basic information on the lizards' appearance and geographic ranges, I avoid highly technical details to keep the vibe of this book light and fun. In the back matter, I refer you to additional resources, including technical field guides. My goal is to teach you how to successfully find, watch, (sometimes) capture, and admire the multitude of lizard species that call California home. Along the way, I will regale you with tales about what makes each lizard species unique and fascinating, and I will feed your eyes with photographic candy from my gifted camera-toting friends. Strap down your pocket watch—you are about to dive into a wonderland of lizards.

A Small-scaled Brush Lizard takes refuge from the sun underneath palm trees.
Photograph by Jeff Lemm.

INTRODUCTION

Lizards in Wonderland:
Why California Has So Many Lizards

If you are a Californian, especially a Southern Californian, you might take lizards for granted. Most kids in suburban areas go through a phase of capturing blue-belly lizards in their yards. Lizards scattering around our feet as we hike on a mountain trail are as much a part of the Californian landscape as oak trees and Red-tailed Hawks. In many Southern California communities, people are so used to the adorable geckos congregating around buildings' outdoor lights that they forget these exotic lizards weren't here until recently.

But lizards are not a given in many other states. When I was twenty years old, I spent a summer in South Dakota doing reptile surveys. We recorded lots of salamanders, toads, and snakes, but the two species of South Dakotan lizards (only two!) remained elusive. When I started asking locals if they'd seen them, they said, "Sure! Spring lizards are a dime a dozen. There are lots of 'em down by the creek." Reader, I searched my heart out for those "spring lizards" week after week, with no luck. I did see plenty of big, beautiful tiger salamanders in the moist areas by the creeks, but no lizards. It finally dawned on me that these "spring lizards" and the tiger salamanders I saw in abundance were actually one and the same. To those locals, four legs and a long tail meant lizard, 300 million years of evolution separating amphibians and reptiles be darned.

Unlike in South Dakota, lizards of the reptilian variety are indeed one of the most common types of wildlife you will see in California. Why are lizards scarce in South Dakota and common in California? While many factors are at play, climate is a major one.

A Southern Alligator Lizard in a meadow, one of many habitats in which this species is found.

Basking lizards like this "blue-belly" (Western Fence Lizard) are pervasive in California.

A Coast Horned Lizard in sandy habitat near the beach. *Photograph by Francesca Heras.*

The single biggest factor determining lizard diversity is temperature: most lizards like it hot. Southern California has a warm climate that affords lizards a long active season to capture prey, mate, and produce offspring. Lengthy winters are much less compatible with reptilian lifestyles. Another factor is the complexity of the environment. Imagine taking a flight from San Diego to Las Vegas. From your window seat, you would see lots of different habitats, from beach dunes and chaparral to concrete urban jungles, transitioning into forested mountains surrounded by lowland scrub, then finally the huge Mojave Desert. Plus, each of these habitats on its own is complex in terms of topography, elevation, plant life, and more. All these factors have facilitated the evolution of many specialized species of life in California—insects, fungi, plants, and just about everything else, including lizards.

A Desert Night Lizard in its typical Joshua tree forest habitat.

There are many other reasons why lizard diversity in California is relatively high, but there is one in particular that we will return to over and over in this book. Southern California is home to a large—and ever increasing—number of non-native species of lizards. As I am wrapping up this book in the fall of 2024, at least fourteen species of non-native lizards can be found in Southern California. This means that one in five species of lizards in California is not native to this state and has been introduced, by people, on purpose or by accident, mostly within recent decades. In fact, after I had already completed a draft of this book and was in the editing phase, a colleague emailed to alert me to a new population of Bosc's Fringe-toed Lizards (*Acanthodactylus boskianus*), native to the Middle East but found in October 2023 in an industrial area in Ventura County. Have we discovered this population in time to capture all the lizards, or will it become a permanent resident like so many others? By the time this book is in your hands, there could be even more established populations and species. Why? In each species account, I discuss the unique story behind each of these tailed invaders' journeys into California. You will see that many of these species occupy suburban residential areas where they thrive in our yards, climb about brazenly on our houses, and take shelter underneath our roof tiles. The word for wild animals that thrive in areas around people is *synanthrope* (*syn* = with or together, *anthro* = people). Southern California, due to its warm climate and dense human populations, is a literal breeding ground for invasive, synanthropic lizard species.

Whether you are enjoying the antics of blue-bellies on an oak tree, searching the skeletons of dead Joshua trees for night lizards, or watching non-native geckos tussle on the outside of your living room window, California is truly a wonderland of lizards. But before we dive down this rabbit hole full of lizards together, let's wrap our heads around the most basic, but surprisingly complicated, question of all: what exactly is a lizard?

What Are Lizards?

As I was writing this book, I taught herpetology for the fifteenth time at Cal Poly. This glorious class introduces students to the biology of amphibians and reptiles, two distantly related groups that are nonetheless smushed into the same course because they share certain characteristics, most notably that they get most of their body heat from the environment instead of from their own metabolism. Also, early taxonomists grouped amphibians and reptiles together for the same reason my friends in South Dakota thought salamanders were lizards: they look kind of similar. Modern molecular techniques have helped scientists tease apart the true evolutionary relationships among these fascinating animals, though the ghosts of taxonomists past remain in some species' names today. To understand what a lizard is—and what it isn't—we need to have a vocabulary lesson. Unlike my herpetology students, you won't be quizzed on these terms, but vocabulary will help you navigate the wide world of lizards.

Starting broadly, lizards are members of a class of animals called the Reptilia, which are vertebrates with scaled skin and with eggs that are encapsuled in membranes to protect the developing embryos. Both characteristics were important adaptations that allowed their ancient ancestors to avoid dehydrating when they moved onto land. Reptiles consist of turtles, crocodilians (crocodiles, alligators, and gharials), and a large group of animals known as the Lepidosauria. This latter group can be further subdivided into two orders, Rhynchocephalia and Squamata. You'll be forgiven if you've never heard of rhynchocephalians, because that group consists of just one species, the Tuatara, a somewhat lizard-like reptile that currently only inhabits islands off New Zealand. On the other hand, you know squamates well, even if not by that name. This incredibly diverse group diverged from ancestors of the Tuatara over 200 million years ago and radiated into the thousands of lizards and snakes that today occupy almost every inch of the globe.

Lizards and snakes? That's right—these animals are very closely related. In fact, over millions of years, multiple different lineages of lizards lost their legs, allowing them to slither more easily through grass or soil to escape predators or to stalk prey. One lineage gave rise to a huge number of species, and we call this lineage the snakes. Another mostly legless lineage, the amphisbaenians, is one you might never have heard of because just one species occurs in the United States (in Florida). There are still other lineages that lack legs, which we collectively call the legless lizards, including

A Desert Spiny Lizard watches the photographer warily from his perch.

a family that is endemic to California and Baja California, Mexico (Anniellidae, see page 45).

But when most of us think of lizards, we conjure an image of a scaly critter with four legs and a long tail. Lizards occupy a huge range of body sizes, from a hatchling chameleon in Madagascar that can perch comfortably on the head of a match, to an enormous Komodo Dragon in Indonesia that can wolf down an entire deer. Even in California, our lizards have a wide range of sizes, from tiny night lizards that spend most of their time hiding under the fallen branches of Joshua trees, to broad-headed, chunky collared lizards that crunch the skulls of smaller lizards before eating them, to huge but elusive Gila Monsters that we occasionally encounter in the far eastern desert borderlands of California. In this book, I broadly categorize California lizards as small (would fit in your palm), medium (would fit in your outstretched hand), or large (bigger than your hand).

Zeev Nitzan Ginsburg.

This might look like a snake, but it is a San Diegan Legless Lizard.

A young Great Basin Collared Lizard awaits an approaching storm.
Photograph by Marisa Ishimatsu.

Scientists have described over seven thousand species of lizards so far, and they blanket the globe, absent only from the polar regions, the tops of the highest peaks, and several oceanic islands. Nearly sixty species from fifteen different families occur in California. Given this diversity, there are no rules that govern California lizards entirely, though I can safely say that *most* of our lizards eat insects, lay eggs, exhibit territorial behavior, and rely a lot on vision when it comes to hunting prey and finding mates. In the species accounts that follow, you will see that much of the physical appearance, ecology, and physiology of Californian lizards revolves around these characteristics, but you will also read about fascinating departures from these rules that will make you look at lizards in a new light.

A Tale of Two Lizards

People love to categorize things into bins. Your snack is sweet or savory. You argue with your friend about whether you should watch a comedy or drama at the movies this weekend. The dress code is casual or formal. However, we all know that there are actually gradations inherent in these categories. The long line at the kettle corn stand every week at our local farmer's market attests to the popularity of a snack that is both sweet and savory, with a nice salty kick, too. Similarly, a phrase that comes up again and again when I teach biology classes is "Biology hates a binary." In other words, as soon as you delve into a topic, you realize that an apparent binary represents two ends of a spectrum, and it is possible to fall anywhere in between. We like to categorize rattlesnakes as ambush foragers and gopher snakes as active hunters, but if you watch them long enough, you'll see that they sometimes stray from these extremes.

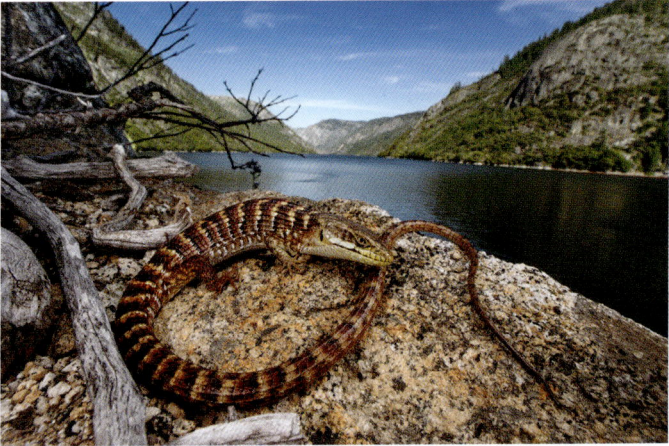

Chad Lane

Southern Alligator Lizards occasionally bask in the sun but mainly lurk in the shade of vegetation.

People apply the binary "nature versus nurture" to ask whether a behavior is genetic and inborn versus caused by the environment, when the answer is invariably "both."

If we understand that most binaries are not set in stone, but rather are useful bins for categorizing things, binaries can be very helpful for learning. In my long history of teaching herpetology, I've come to rely on certain binaries to describe the biology of lizards and other creatures to my students. Bins help students memorize, understand, and conceptualize, an important first step in learning the vast world of herpetology. Once they are equipped with these concepts, it's time to highlight the departures from the rules and note the fact that many of these binaries represent two ends of a continuum, with real-life animals falling somewhere in between.

With that in mind, let's dive into what I like to call "A Tale of Two Lizards." We are going to examine the biology of California lizards by binning them into several binary categories, which I will then refer to again and again in the individual species accounts.

Nocturnal lizards, like this Peninsula Leaf-toed Gecko, are concentrated in the deserts of Southern California where nights are warm. *Photograph by Bryce Anderson.*

1. Diurnal vs. nocturnal

Most California lizards are active during the day (diurnal). Whether they bask in the sun or move around underneath or even within a cover object like a log, it is common for lizards to go about their business when temperatures are higher. One reason for this is their size. Most California lizards are very small, which gives them a relatively high surface area compared to the size of their bodies. This means that they warm up and cool down quickly, since heat can easily move across the high surface areas of their bodies. Being active during the day gives lizards access to enough heat for them to thermoregulate effectively, often by moving between warm and cool areas to keep their body temperatures right at the sweet spot. On the other hand, California has several species of lizards that are active exclusively at night (nocturnal). These include

Chad Lane

Many diurnal California lizards, including this Common Chuckwalla, bask in the sun to regulate their temperature, making them rather conspicuous and easy to spot.

certain desert lizards like the Western Banded Gecko (see page 77) and the Peninsula Leaf-toed Gecko (see page 150), which emerge from their hiding places after sunset to stalk about the desert in search of prey. Deserts are often warm at night during the spring and summer, which facilitates thermoregulation of these little lizards. California also has populations of several species of non-native geckos, all of which are nocturnal and live mostly on man-made structures like buildings. While the diurnal-nocturnal binary is very useful for helping you learn when to search for lizard species, you might find a lizard breaking the rules now and then, especially when it is unusually cool or warm out.

2. Baskers vs. lurkers

Many species of lizards bask in the sun to thermoregulate, to maintain proper mineral balance in their blood and bones, and for other

Marisa Ishimatsu

Most Californian lizards have color patterns that help them blend in with the background, like this Mohave Fringe-toed Lizard partially buried in sand.

physiological reasons. These lizards are usually highly conspicuous and familiar to people, especially the ubiquitous blue-belly (Western Fence Lizard) which basks on just about any surface throughout most of our state. But many other species avoid the sun during the heat of the day, or even shun it entirely, instead thermoregulating by moving about underneath vegetation or cover objects. Alligator lizards and skinks, for example, might be seen out in the sun occasionally, but you are far more likely to find them lurking about in the shade or hiding underneath a piece of tin or a fallen branch. In this book, each species account has a section describing how to search for that species. Knowing whether the species is a basker or a lurker will help you know where to look for it and how to form a search image—an image in your mind of the lizard in its habitat that helps you find it in the wild. However, there are cases where species depart from this binary. Whiptail lizards employ both strategies by constantly staying on the move; at any moment, you might find them scrabbling around in leaf litter in the shade or pausing in the sun to soak up rays. Nocturnal lizards by definition don't bask in the sun, but they don't necessarily skulk around under vegetation, either. Over time, as you read the species accounts and go lizard hunting, you will become familiar with each species' habits and form your own understanding of the basker-lurker continuum.

3. Colorful vs. camouflaged

When it comes to lizards and the landscape, some blend in against their background while others are screaming to be seen. Why is that? It's complicated. As with other reptiles, and indeed animals in general, camouflage provides the obvious benefit of allowing critters to blend in with their environments so they are less likely to be eaten by a predator. Flipping through the photos in this book, you will notice that many Californian lizards are some shade of brown or gray, with patterns that help them blend into trees, dirt, vegetation, rocks, or other substrates. Being colorful is the exception to the rule of camouflage in lizards, but even bright colors may not mean what you think they do. The mighty Gila Monster has bright pinkish-orange wavy bands alternating with black, which may be an adaptation to warn predators that this lizard is venomous. But this is not as straightforward as it seems. When radio-tracking Gila Monsters in Arizona during graduate school, I noticed how hard it was to see these big lizards when they were in the mottled shade of vegetation. However, in most lizards, the main function of bright colors is to communicate with other lizards. For example, male lizards of many species in the family Phrynosomatidae have vivid colors on their undersides that intensify during the mating season. In some species, like the Granite Spiny Lizard, the deep blue can extend from his snout and fingers to his toes, and even his back can

Lee Grismer

Western Fence Lizards blend into the background from above but show off their bright-blue bellies to communicate with rivals and potential mates.

be iridescent purple, orange, and yellow. Are you ready to have your mind blown? It turns out that the colors we see on lizards are only a fraction of what is actually there. Our human eyes can see certain wavelengths of light, from the short wavelengths that we see as purple to the long wavelengths that represent red. Diurnal lizards are capable of seeing those colors, but they can also detect even shorter wavelengths that reflect ultraviolet light, such that what might look to us like a plain white chin on a male lizard actually appears to them as a blazing ultraviolet signal they use to "talk" to one another. They can also see long wavelengths in the infrared, but we don't know much at all about this sense. Hopefully you can now see what I mean when I say that lizard color is "complicated."

4. Egg laying vs. live bearing

The vast majority of California lizards lay eggs to reproduce, just like you would expect them to. However, viviparity (live bearing) has evolved independently numerous times among the lizards of

Jeff Martineau

Mediterranean House Geckos were the first non-native lizards introduced to California, where they are now common on buildings in many neighborhoods.

the world, just like leglessness has. There are a few examples of live-bearing lizards in California, including our legless lizards and night lizards, certain species of alligator lizards and horned lizards, and some of the non-native species that have been introduced into California. Why would live bearing evolve? One hypothesis involves the ever-important theme of thermoregulation. Instead of laying eggs that could be impacted by low or highly variable temperatures, the lizards keep the eggs inside their oviducts, don't bother to produce eggshells, and carry the developing embryos around in their bodies until they are born as live young. This allows the pregnant females to keep the developing babies cooking at the perfect oven temperature by basking in the sun or retreating into shade. This also protects the eggs from notorious egg-devouring predators. There is always a trade-off, though. Live-bearing lizards have to drag around the extra weight of the embryos for weeks or even months, making it harder to find prey and to avoid predators.

5. Native vs. introduced

About 80 percent of the lizards in this book are native to California, but the others have been introduced by people. As I mentioned, many of these species have taken root in the hospitable climes of Southern California and are here to stay, while for some newcomers, scientists are working feverishly to try to eradicate them before they can become permanently established. It might seem like an insurmountable task to find *all* the tiny lizards in such an effort, but the dedication of the scientists that search for these proverbial needles in a haystack can win out. In 2018, scientists discovered wild African Five-lined Skinks in Glendora, California, that had apparently escaped from a local person who imported reptiles for the pet trade. Through a Herculean effort that involved enlisting the whole neighborhood as community scientists to support eradication efforts, this invasion was quashed before it could take permanent hold. Importantly, scientists found out about the skinks via a post on iNaturalist (a website and app where people post photos of organisms they encounter), highlighting the central importance of collaborating with community scientists who take the time to photograph wildlife in their neighborhoods and post photos to this important platform. If you are bitten by the lizard-hunting bug, I highly encourage you to make your own iNaturalist account and start contributing photos, especially if you see lizards in residential areas that you cannot identify using this book or a field guide.

Hopefully my "Tale of Two Lizards" has provided some pointers that will help ensure that your lizard-hunting trips fall into the category of the best of times rather than the worst of times. After all, a day wasted on lizards is not wasted on one's self. It's time to set Dickensian puns aside and take a closer look at how to search for and observe lizards in California.

How to Find and Watch Lizards in California

If you read my book *California Snakes and How to Find Them*, you will know that I devoted a detailed section to the art of finding sneaky serpents. Indeed, finding those legless escape artists requires a lot of skill, luck, and persistence. Luckily, lizards are overall easier to find than snakes. This is especially true for the diurnal, sun-basking lizards that are so common throughout California, where you can see dozens of them when you hike through a natural area or walk lazily around your neighborhood. However, even these baskers follow seasonal and daily patterns, and many of the lurkers can be challenging or sometimes feel next-to-impossible to find. As with any animal, a little bit of knowledge about California lizards will go a long way in helping you find them.

As we have learned, most Californian lizards are active during the day, and many of them are baskers. For these species, as a general rule, you should go search for them on sunny days with little wind, when the temperature is about right for wearing a T-shirt. In the deserts, some lizards might prefer temperatures a bit higher—by the time Desert Iguanas come out to forage, you will wish you had a swimming pool to jump into. By and large, the few native nocturnal lizards of California occupy warm areas like the inland deserts, and many of the nocturnal introduced species live on buildings in Southern California where they are buffered from low temperatures.

Hiking is the best way to find most basking lizards, whether it be on a trail through the rocky desert in search of a spiny lizard or on sand dunes hoping to catch a glimpse of a speedy fringe-toed lizard. Walk slowly and quietly, keeping your eyes trained on the habitat. In the beginning, you'll see these lizards out of the corner of your eye as they dash away from you because they spotted you

Some lizards are widespread and conspicuous, while others (like this Granite Night Lizard) have specific habitat requirements and are camouflaged.
Photograph by Zeev Nitzan Ginsburg.

first. But after a little while, you'll develop a search image for them. You will be able to detect the profile of a lizard's head peering at you from a distant rock. You will notice a smudge in the sand that doesn't match the rest of the dune because it's actually a partially buried lizard. Hiking can also take you to the best places for lurkers, but you will need to employ different tactics to find them. These might include covering ground quickly hoping to flush these lizards (i.e., cause them to run from their hiding spot), flipping cover items to look for lizards hiding underneath, and more. Both baskers and lurkers can seek shelter from extreme weather in rock crevices, inside burrows, within plants, or under cover objects. The optimal searching method depends on the species, and I discuss strategies in each species account.

Finding lizards is just the first half of the battle. Next, you need to watch them. Please!! Grab some binoculars and perhaps a camera with a zoom lens and go out on some lizard-watching hikes. Resist

Two Common Chuckwallas mating.

the urge to capture them (at least for a little while). Why? Because lizards do cool stuff. Many lizards are territorial, meaning that they actively defend a spot (a tree, rock, section of fence or building, or just a patch of the ground) against competitors. They are defending a resource—sometimes food but often mates. This means that you can regularly see lizards squabbling over a choice perch, often in fascinating or hilarious ways. Because they can't shout at one another, they use exaggerated behaviors to communicate visually. They circle one another like tiny sumo wrestlers. They jut out their colorful chins at each other as they flatten their bodies sideways to appear bigger than their opponent. Sometimes they even go in for the attack, biting Mike Tyson–style at their adversary's head.

You might also get to see the crème de la crème of behaviors— lizards engaging in courtship or mating. If you see one lizard following another around relentlessly, it's probably a male interested in a female. Sometimes he bobs his head up and down at her during his pursuit. He might wrap his tail around hers in hopes that she'll accept him as a mate. Often, he'll grab her by the nape of the neck with his mouth as they mate.

You might also get lucky and see a lizard chasing down its dinner. I've seen Western Fence Lizards launch themselves from their perches to catch a fly or a bee in midair. I've spent the better part of a morning skulking in the wake of a Gila Monster as it poked its head into burrows in search of rodent nests to raid. I've giggled uncontrollably after encountering a Long-nosed Leopard Lizard that had eaten a whiptail lizard so large that the long tail still stuck out of its mouth. Likewise, you might witness a lizard becoming a meal for a hungry snake or other predator. As my students say, "Nature is metal."

But my favorite example of cool-things-you-could-see-if-you-watch-lizards wasn't something I saw, but a fascinating interaction between baskers and lurkers witnessed by my friends Mary and Bill. I wouldn't have believed this without photo documentation, but they were prepared with their camera and a lot of patience as they watched the drama play out. They saw a pair of Western Whiptails (lurkers) digging up a nest of lizard eggs and eating them one by one until they were confronted by an angry Western Fence Lizard

Marisa Ishimatsu

A Sidewinder swallows a Desert Iguana.

(the basker and presumed parent of the eggs). This was a fascinating observation for many reasons, the biggest one being that fence lizards aren't typically thought to defend their nests against predators. Mary and Bill became community scientists who added an important tidbit of knowledge to our natural history treasure chest, all because they took the time to watch these common lizards instead of walking blindly by them or disturbing them by catching them.

This Western Whiptail raided a Western Fence Lizard's nest to eat the eggs, until the fence lizard confronted it and scared it away. *Photographs by Bill Walker.*

Lassoing lizards is an essential skill for those planning to capture and admire lizards.
Photograph by Brittany App.

Catching Lizards: A Cautionary "Tail"

We've discussed why California is so lizard-rich. We have had a vocabulary lesson about what makes a lizard a lizard. We've delved deeply into some useful lizard binaries. Then we talked about how to find and watch lizards in the hopes of seeing some of their legendarily exciting behaviors.

Now, let's talk about catching lizards. Before we jump in, let me reiterate that *not catching* lizards can be just as fun as catching them because you can watch them do amazing things that you would miss out on if you grabbed them right away. But that is not the only reason to think twice before chasing down every lizard you see. Being captured and held by a person is stressful to a lizard. Many species of lizards have the fascinating ability to drop their tails off when handled by a lizard hunter, whether that's a predator or an over-eager person (hence the pun in the title of this section . . . did you get it?). The newly liberated tail wiggles around and distracts the predator, sometimes allowing the lizard to get away. Yes, they can regrow their tails, but this costs them lots of energy, so it is best avoided.

Some people believe that we should never handle wildlife without a good reason, and others have different value systems and simply try to minimize stress by being careful when capturing lizards and letting them go shortly after catching them. My students catch lizards to examine them up close and learn about them, and sometimes as a necessary part of collecting data for a research project. There truly is no substitute for holding a lizard up to your eyes to see the intricate patterns on its skin, its reinforced scales, its curly toes. I implore you to seek balance when catching wildlife: always be gentle, avoid grabbing lizards by their tails, hold them for

This newborn Pygmy Short-horned Lizard is so adorable that you can understand why many people want to catch lizards to admire them up close.

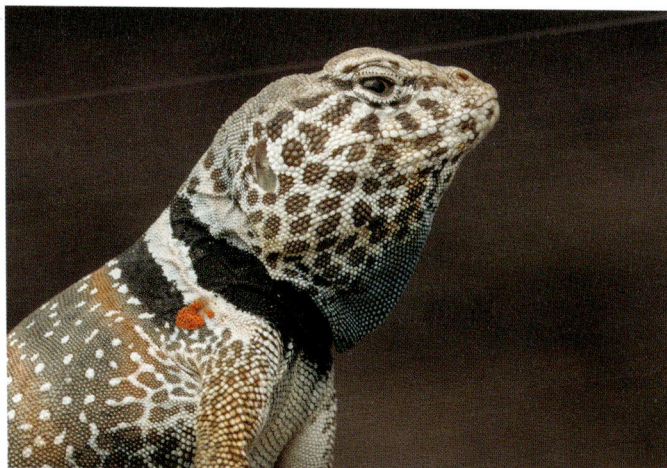

Observing the intricate details of features like this Great Basin Collared Lizard's scales and the colony of orange mites on his shoulder requires catching a lizard (or zooming in with an excellent camera).

as short a time as possible, and follow all local, state, and federal regulations.

I have witnessed generation after generation of students being bitten by the lizard-catching bug after heading outside with a lizard pole and holding their prizes for the first time. In many areas, lizards are super abundant, so you can get lots of practice rapidly. However, each species is different, and they all require unique strategies to get them in hand. I will give you some tips here that you can try, and then you can hone your own skills to perfection. But first, here are a few things to think about before you grab your pole.

Rules and regulations

As with all wildlife, you are not allowed to touch or otherwise interfere with lizards in state and national parks as well as in some other parks and reserves. In other areas of California, you are permitted to catch and release lizards and other reptiles if you hold a current fishing license from the California Department of Fish and Wildlife. In 2024, the license ran about $60 for California residents and $165 for visitors from out of state. The license also allows you to collect a certain number of lizards to keep as pets, although I think that wild lizards should stay in the wild. If you want a pet, consider buying a Leopard Gecko or another cute, captive-bred lizard at a reptile expo.

Flipping cover objects

Just as snake hunters know to look for those creatures hiding under cover objects on the ground—like pieces of tin, wood, cardboard, rocks, even old appliances—seasoned lizard hunters know a good rock to flip when they see one. You can find many of the lurkers under cover objects, including alligator lizards, legless lizards, skinks, night lizards, and some geckos. Even baskers will hide under rocks or other cover objects when it is too cold or too hot outside for them. Be sure to follow the golden rule when you flip cover objects: replace all objects in the exact same spot you found them so that the critters' homes are not ruined (for them and for future lizard hunters). Finally, never stick your fingers under a piece of wood or plunge your hand into a burrow, because venomous animals like scorpions and rattlesnakes may be hiding within.

Lasso that lizard!

While many lizards can simply be picked up by hand, others won't let you get close enough for that. Perhaps the most essential tool for the lizard hunter is the pole and lasso. The idea here is that a lizard will run for cover if a person walks right up to them, but they don't recognize a pole with a lasso on the end of it as a threat. So, a person standing ten feet away can slowly extend their pole toward

This endangered Blunt-nosed Leopard Lizard is being lassoed to collect data as part of a permitted research study. *Photograph by Brittany App.*

the lizard, place the lasso around its neck, and then tug the lasso tight to capture the lizard. These contraptions come in many forms, ranging from makeshift devices like a stick with a loop of grass tied to it (my friend used strands of her hair when she was a kid!), to high-tech fishing rods outfitted with the hunter's lasso material of choice. Some people like to use dental floss, some use silk thread, and others use fine string. I prefer braided fishing line tied on to a ten-foot telescoping crappie rod, both available at your local fishing store or online. There are many ways to tie a lasso, which you can figure out by trial and error or by looking it up online. Successfully lassoing lizards takes practice, but watch out! It can be addictive.

How to hold a lizard

The whole point of catching a lizard is to hold it in your hand and admire it. If you lasso a lizard, you'll need to hold it to remove the lasso from around its neck. So, if you're going to catch a lizard, you need to learn to hold it properly. There are several considerations when handling a lizard. First, to avoid them dropping their tails, never grab them by the tail. Next, some lizards will bite you if you let them. Though bites from all Californian lizards are harmless (except the exceedingly rare venomous Gila Monster), some toothy species can pack a punch and you'd probably prefer to keep your skin intact. Grip lizards in such a way that they can't whip around and bite you. You also want to make sure you are holding them with a gentle and comfortable grip. Acceptable grips include gently wrapping your fist around the lizard's body with its head emerging between your thumb and index finger; however, lizards could overheat this way, so be sure to let them go rapidly. A better grip, which also allows others to admire the entirety of the lizard's body, is to hold a front leg and a rear leg between your thumb and hand, as shown in the facing photo.

Nicolette Murphey

A safe and gentle way to grip a lizard is by its front and hind legs. This endangered Blunt-nosed Leopard Lizard was handled as part of a permitted research study.

Last but not least, be sure to keep lizards out of the sun when you hold them. Even though baskers are regularly found in the sun, they maintain a proper body temperature by shuttling back and forth between the sun and shade and can overheat rapidly if forced to stay in the sun. This is even more important for lurkers or nocturnal species like night lizards and geckos; if you find them under a cover object during the day, take care to keep them completely out of the sun.

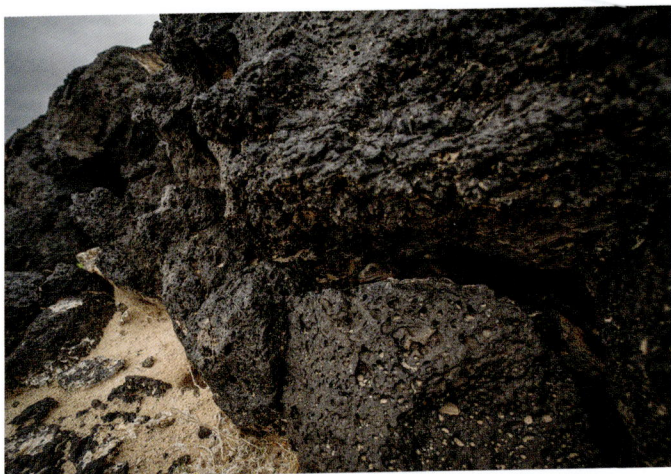

Common Chuckwallas retreat to rocky crevices to hide from predators and eager lizard observers.

Once you have sufficiently examined and admired your prize, be sure to let the lizard go in the exact spot where you captured it. After it shakes off its fear from being held by a big, hairy primate, you will find that it often rapidly crawls back under its cover object or returns to its perch and resumes its lizardly ways.

You are now equipped to head out for a day of lizard hunting! Pack your water bottle, hat, binoculars, camera, this book, plus your fishing license and lizard pole if you plan to catch and release them. Head out to a local park or other destination with wide open natural areas to explore. Stroll across a desert sand dune watching for the pale lizards scattering in all directions. Climb around on rocks looking for lizards hiding in crevices. Above all else, keep your eyes open and your attention focused on our (mostly) four-legged friends. Prepare to lose your sauroblivion as the world of scaly little creatures opens up before you.

Coast Horned Lizard. *Photograph by Spencer Riffle.*

THE LIZARDS

NORTHERN ALLIGATOR LIZARD

ELGARIA COERULEA

FAMILY ANGUIDAE

Alligator lizards are good examples of lurkers, in that they typically hang out underneath objects or move around in the shade of vegetation, rarely sitting out in plain sight. For this reason, they are not as obvious as other species that occur where they do, like the Western Fence Lizard. However, alligator lizards can still be common in certain areas. When I was an undergraduate at UC Berkeley, I volunteered to help a graduate student named Judy collect alligator lizards for her study. She needed both Southern and Northern Alligator Lizards, which occur in the same areas in some parts of the state. We drove up the coast from the Bay Area, pulling out at beachside rest areas and hunting for alligator lizards lurking amid the vegetation. We found lots of them when conditions were best: foggy, still mornings or afternoons when the ground was surprisingly warm. The lore that lizards are only out when it is sunny is simply not true, especially for lurkers like these. As long as it's not too cold, the cloudy or foggy conditions are perfect for them to thermoregulate under the vegetation and ideal for them to hunt the bugs they eat. Alligator lizards are fierce little guys, especially the males, whose heads grow wide and strong under the influence of testosterone. All the better to bite other males vying for the same female, or to clamp down on the fingers of the inexperienced students who catch them. I certainly got my share of nips from alligator lizards on this trip. If you try to catch them, watch out not only for their bites (they can break the skin!), but also be sure not to grab alligator lizards by their tails, which will break right off as a tactic to get away from predators.

Appearance: Northern Alligator Lizards got their name because their thick, armored heads make them look somewhat like an alligator. However, lots of people actually mistake them for snakes because their bodies are elongate and they have tiny limbs. These medium- to large-sized lizards have robust bodies and triangular heads. Their color is highly variable, often brown or gray with white spots that can give an irregular patterned appearance, and cream or yellow underneath. They have really long tails, which they can regrow if they break off. Regrown tails typically lack a pattern, so it is easy to tell whether a lizard has regrown its tail. Their bellies can help you differentiate them from the closely related Southern Alligator Lizard: both have bellies covered with a series of parallel, rectangular scales, and in the Northern

Alligator Lizard there are dark stripes that run lengthwise *between* the scales, while in the Southern Alligator Lizard the stripes run through the *middle* of the scales.

Natural History: In addition to the coastal areas described above, Northern Alligator Lizards can be found in grassland and woodland habitats. They are most common in areas where they have plenty of access to logs to hide under as well as bushes that provide shade, and they are less common in thick forests with little sunlight. They climb around on low vegetation readily and can also swim when they need to get away from a threat. They eat all types of invertebrates, and large lizards occasionally eat small vertebrates. They mate in the spring, and females do not lay eggs but instead give birth to live young in the summer.

Range and Variations: Northern Alligator Lizards occur in coastal counties from the Monterey Bay northward, then through much of Northern California, and in the Sierra Nevada as far south as Kern County. They also extend northward through western Oregon and Washington into southern British Columbia, and into small areas of Idaho and Montana. There are four subspecies of Northern Alligator Lizards, all of which occur in California.

How to Find Northern Alligator Lizards: Northern Alligator Lizards can be found by hiking through appropriate habitat and watching for the lizards lurking under vegetation. Sometimes one will see you before you see it, and you will accidentally flush it. They usually stay above ground, so you can slowly approach and see if you can find it hiding amid the brush. Flipping cover objects is another excellent way to find Northern Alligator Lizards. They can

Chad Lane

be found underneath almost anything, though I have the best luck finding them underneath wood, including driftwood near beaches, logs in woodland, or plywood placed out on purpose to attract reptiles and other critters.

Chad Lane

SOUTHERN ALLIGATOR LIZARD

ELGARIA MULTICARINATA

FAMILY ANGUIDAE

If you live in Southern California, you might have seen an article in the newspaper or on your social media feed around Valentine's Day each year asking you to be on the lookout for "Lizards in Love." The article instructs you to report any sightings of courting Southern Alligator Lizards to iNaturalist or directly to researchers at the Natural History Museum of Los Angeles County who are studying alligator lizard reproduction. What does courtship look like in Southern Alligator Lizards? Alas, there are no miniature bouquets or boxes of chocolate being gifted, nor is there any holding of tiny, scaly hands. Instead, a male alligator lizard bites onto the nape of a female's neck while waiting until she acquiesces to mating with him. These "love bites" can last for hours or even days. By asking community scientists (like you!) to report observations, the scientists have learned far more about alligator lizard reproduction than they would have on their own, showing that science truly is a collaborative endeavor between scientists and members of the public. Images of Southern Alligator Lizards posted by community scientists have revealed additional secrets as well. These lizards occur in people's yards and in parks in the Los Angeles metropolitan area, and one study analyzing photos uploaded to iNaturalist showed that lizards are more likely to experience predation attempts in urban areas, as evidenced by the greater frequency of regrown tails. Though these lizards had gotten away, many do not. The probable culprits are outdoor house cats, which kill *billions* (yes, billions!) of wild animals each year. For the sake of alligator lizards and loads of other wildlife species, keep your cats indoors.

Appearance: Southern Alligator Lizards are medium to large in size, with their bodies about as long as your hand and their tails extending much farther. They have triangle-shaped heads that are larger and wider in males. They usually have mottled colors on their backs, often brown, yellow, or reddish. Hatchlings are often solid-colored and only later develop the mottled pattern. The definitive way to distinguish Southern from Northern Alligator Lizards is from their belly pattern: both have bellies covered with a series of parallel, rectangular scales, and in the Southern Alligator Lizard the stripes run through the *middle* of the scales, while in

Max Roberts

the Northern Alligator Lizard there are dark, lengthwise stripes that run *between* the scales. In addition, Southern Alligator Lizards usually have lighter-colored eyes.

Natural History: Southern Alligator Lizards inhabit grassland, oak woodland, chaparral, and yards in urban areas. They are most active during the mornings and evenings when temperatures are mild. When active they lurk in the shade of bushes and only occasionally venture into the sunlight to warm up. Like all species of alligator lizards, they spend a lot of time hiding under cover objects. They eat insects and other invertebrates, with the occasional small vertebrate consumed by particularly large lizards. Southern Alligator Lizards mate in spring, lay eggs in summer, and hatchlings appear in late summer.

Range and Variations: Southern Alligator Lizards occur throughout the entire western half of California as well as the Sierra Nevada and Owens Valley. They are absent from the eastern deserts and the Central Valley. Outside of California, they extend north into southern Washington and south into northwestern Baja California. There are two subspecies of Southern Alligator Lizards, with the transition occurring in the Monterey Bay area.

How to Find Southern Alligator Lizards: The easiest way to find Southern Alligator Lizards is to flip cover objects in good habitat, especially logs and junk like plywood or tin. You can also find Southern Alligator Lizards active above ground during times with mild temperatures. Look for them lurking under bushes. You are unlikely to find these lizards basking in direct sunlight.

PANAMINT ALLIGATOR LIZARD

ELGARIA PANAMINTINA

FAMILY ANGUIDAE

Panamint Alligator Lizards are a truly Californian species. They were only discovered about sixty years ago in their habitat at mid-elevation within several mountain ranges in eastern California. These lizards are more closely related to Southern Alligator Lizards than they are to Northerns. They historically occurred in only a very limited geographic area, and so pressures from humans are having a disproportionate impact on Panamint Alligator Lizards. You might think that mountains towering above our eastern deserts are remote areas with little human impact, but think again. The area is riddled with thousands of mines, including for gold, silver, lead, and numerous other elements. In addition, the area has become popular recently with off-road vehicles. These stressors have impacted Panamint Alligator Lizards enough for the California Department of Fish and Wildlife to label them as a species of special concern.

Appearance: Panamint Alligator Lizards are medium-sized lizards with tails that can be twice as long as their bodies. Adults have a banded appearance, with tan, gray, or yellowish bands alternating with darker brown ones. Juveniles have especially high-contrast bands. Panamint Alligator Lizards have large, triangular heads, and their eyes are typically pale around dark pupils.

Natural History: Panamint Alligator Lizards occur in grasslands, scrub woodland, and moist areas near creeks at mid-elevations, and they are absent from dry desert areas at lower elevations. They are typically active during the day and are only seen at night when it is very hot out. Areas with plentiful brush or with rocky outcrops are used for cover. Like other species of alligator lizards, they eat invertebrates. Their reproduction has not been studied in detail, but they mate in spring and likely lay eggs in the summer, with juveniles hatching later in the summer or in fall.

Range and Variations: The Panamint Alligator Lizard occurs at around 2,500–7,500 feet on mountains throughout Inyo County and southern Mono County. There are no subspecies.

How to Find Panamint Alligator Lizards: Panamint Alligator Lizards are not tremendously easy to find in most of their range because they occur in rugged areas off the beaten path. Your best bet for finding them is to search near wet areas like creeks and seeps and in areas with dense piles of small rocks during late spring or summer at times of day when the temperatures are comfortable for you to be wearing a T-shirt. Their protected status means you should avoid capturing Panamint Alligator Lizards.

Clockwise from top: Temblor Legless Lizard, Big Spring Legless Lizard, Bakersfield Legless Lizard, Northern Legless Lizard, San Diegan Legless Lizard.
Photographs by Chad Lane.

LEGLESS LIZARDS

ANNIELLA ALEXANDERAE, A. CAMPI, A. GRINNELLI,
A. PULCHRA, AND *A. STEBBINSI*

~~~~~~

FAMILY ANNIELLIDAE

A legless lizard? Isn't that just . . . a snake? The answer is a resounding no. As I wrote in the introduction to this book, while snakes indeed originated from a lizard ancestor over many thousands of generations, evolving into thousands of new species, this lineage does not encompass the legless lizards. These lizards, including the five species of California legless lizards, comprise one of many other lineages that independently lost their legs over evolutionary time. But these lizards still have other lizardy things that snakes don't have, like eyelids. Why all the leg-losing? The most common hypothesis to explain the repeated, independent loss of legs among lizards around the world is that it made it easier for them to slink through vegetation and burrow into soils. Just take a look at some other lizard species in California to be convinced: many species of skinks and alligator lizards have elongate bodies and tiny limbs and live in habitats where they slink around like snakes. We are possibly seeing these lineages in a snapshot in time, on their way to losing limbs!

Appearance: You're liable to mistake legless lizards for a snake at first glance. However, looking at them closely, you will see that they have eyelids, which snakes lack. They are small lizards with smooth, shiny scales and pointy snouts to facilitate digging into the soil. Legless lizards in California vary geographically in color, with their backs often silver, black, or brown and their bellies typically yellow and sometimes cream. See Range and Variations for more information on the color patterns of the five individual species.

Natural History: Legless lizards in California are burrowing species, spending most of their time underground. As a result, they are typically found in habitats with loose, moist soils with plants, including beachside sand dunes, washes, and sandy areas in woodlands. These lizards eat mainly larval and pupal invertebrates. They mate in the spring and summer and give live birth to a small litter of babies in the fall.

Range and Variations: Up until recently, there was only one recognized species of legless lizard in California, with its closest living ancestors being the alligator lizards. However, scientists studied the DNA of the legless lizards in 2013 and discovered that there were five lineages that are genetically distinct enough to be named their own species. Two of these are widespread and the other three occur in tiny areas. Northern Legless Lizards (*Anniella pulchra*) are highly variable in color and occur from the Bay Area south to Santa Barbara and Los Angeles Counties and are absent from the Central Valley and deserts. The San Diegan Legless Lizard (*Anniella stebbinsi*) is brownish or silvery on top with yellow sides and belly, and ranges throughout coastal Southern California from Los Angeles County southward into northern Baja California, plus several disjunct populations in the southern Sierra Nevada in Kern County. The Temblor Legless Lizard (*Anniella alexanderae*) has a gray belly and black stripes along its back and sides and is present only in a tiny area in the southern part of the Temblor mountain range. The Bakersfield Legless Lizard (*Anniella grinnelli*) is gray with orange sides, a purple belly, and several black stripes down its back and sides, and occurs only in the Carrizo Plain and southern San Joaquin Valley, where much of its habitat continues to be destroyed for agriculture, oil, and housing developments. The Big Spring Legless Lizard (*Anniella campi*) is grayish yellow on top and bright yellow on its sides and belly, and so far, is known to live in several areas with moist soils along the western edge of the Mojave Desert in Inyo and Kern Counties. It is possible that additional populations of these secretive species will be found after further study.

How to Find Legless Lizards: Legless lizards can be found by flipping cover objects in areas with sandy soils and plenty of plants. People living alongside coastal dunes readily encounter them in their yards while gardening. They are also common within the leaf litter underneath oak and other trees. If you are trying to find these animals, you could invest in a potato rake, which you can scrape around in the sandy soil and leaf litter underneath trees. I have had more luck finding legless lizards under flat rocks.

Marisa Ishimatsu

A male Green Anole extends his dewlap to communicate with neighbors.

GREEN ANOLE

ANOLIS CAROLINENSIS

FAMILY ANOLIDAE

Green Anoles are not native to California, but there are several known established populations (and probably many more) that likely arose from the pet and nursery trades. Sometimes called "chameleons" due to their ability to change color, but not true chameleons, Green Anoles are popular pets because they cost only about five to ten dollars and are for sale in most pet stores. However, sometimes they escape, or people let them go in their yards when they no longer want them. (Do *not* let pets go in the wild, as this is how invasive species become established!) It is also possible that these lizards spread to parts of California when their eggs were transported in the soil of nursery plants, given that their native range includes southeastern states like Florida, where many plants are grown and distributed. This species of lizard is also a popular species for studies of endocrinology and neuroscience, but it is unlikely that the wild populations come from escaped research animals. At any rate, Green Anoles now form part of the large and growing group of non-native, introduced lizard species occupying California. Extirpating an introduced species is difficult or even impossible once it has become established, so these lizards are likely here to stay. There is some evidence that Green Anoles, along with other introduced species, may be having negative impacts on native lizards due to competition for food or basking space. You can help scientists study the impacts of non-native species by posting photos of them to iNaturalist.

Appearance: Green Anoles are small, thin lizards with pointy snouts and long tails. They are often, but not always, green. They can change color rapidly according to temperature or their mood, so they might be brown, or green, or both. They often have a light stripe down the center of their back. Male anoles have a pinkish flap of skin on their chin called a dewlap that they can fan out and back in a mating display directed at females, or to taunt rival males.

Natural History: In California, Green Anoles mainly occupy parks and residential areas, where they are active during the day. They are usually found climbing on plants, including tropical plants used for landscaping in these areas. They eat small insects and occasionally vegetation. Green Anoles are prolific reproducers, mating and laying eggs from spring through late summer, often in multiple clutches. Hatchlings appear during the summer and fall. Green Anole activity declines in winter, but they can remain active throughout the year in Southern California areas with mild winters.

Green Anoles can change color and appear brown at times.

Range and Variations: The native range of Green Anoles extends from central Texas eastward to Florida. Due to the pet trade, these lizards have been introduced in many places, including the northeastern United States, Hawaiʻi , Mexico, and several Pacific islands. In California, Green Anoles have established populations in Balboa Park in San Diego County and in Hancock Park in Los Angeles County. In addition, there are numerous other records of Green Anoles throughout coastal Southern California, plus individuals found in Bakersfield, the Bay Area, and Eureka. Whether these latter individuals represent established populations or are just recently released pets is not known.

How to Find Green Anoles: Green Anoles climb on vegetation, where they often blend in and can be difficult to see. However, if you carefully observe your surroundings as you walk, you might see one crawling up a branch or running across the path in front of you. Given that Green Anoles are not native to California, their densities are spotty in most areas where they have been recorded. They are common in Balboa Park near the San Diego Zoo, so that is a great option if you want to see them in the wild. Green Anoles exhibit hilarious and entertaining behaviors to one another, so if you find a good spot, I encourage you to sit and watch them with binoculars. In addition to the dewlap extension described above, male Green Anoles perform adorable head-bobbing and push-up displays at other males invading their territories or to females with whom they want to mate. Sometimes male contests can escalate into all-out brawls complete with wrestling and biting.

William Flaxington

BROWN ANOLE

ANOLIS SAGREI

Brown Anoles appear to be more recently introduced to California than the Green Anole, with the first population of Brown Anoles being documented less than ten years ago. Since then, these anoles have been found throughout the state, though it is unclear how many of these are from established populations. Most of them were likely stowaways in shipments of palm trees and other tropical plants from Florida and Hawai'i, both of which are areas where the Brown Anole is also introduced and is thriving. Brown Anoles therefore are not as easy to find in California as Green Anoles, but documenting the presence of both species on iNaturalist is important to help scientists understand how they are spreading.

Appearance: Brown Anoles are small lizards that are typically brown or gray, and their color can change based on conditions. They have wide heads, narrow bodies, and long tails, though their heads are proportionally not as large as those of Green Anoles. Note that Green Anoles can be green or brown, but Brown Anoles can only be brown or gray. Sometimes they have a series of light spots on the back that form a stripe-like pattern. Males have a dewlap (flap of skin on the chin), usually orange surrounded by yellow, that they extend to signal other lizards. Sometimes males have a crest extending down their heads and backs that makes them appear rather macho.

Natural History: Like Green Anoles, Brown Anoles are also found in California in residential areas with tropical plants and other non-native vegetation. While both may climb on vegetation, Brown Anoles tend to spend more time on low vegetation, on rocks, and on the ground than Green Anoles. They are active during the day and like to bask in the sun. Brown Anoles eat a variety of insects as prey, and large individuals also eat small lizards like hatchling Green Anoles. Though little is known about Brown Anole reproduction in California, it is likely similar to elsewhere in their range. Mating occurs in the spring and early summer, and females lay eggs during the summer. Interestingly, they lay one egg at a time, every two weeks or so, for much of the summer. Hatchlings come out later in the summer and fall.

Range and Variations: The native range of Brown Anoles includes the Bahamas and Cuba. They apparently spread to Florida back in the 1950s, possibly as stowaways on ships. Since that time, Brown Anoles have been reported from every continent except Australia. In California, records are concentrated in Orange County, with other observations in surrounding southern counties as well as single individuals in Sacramento and Bakersfield. They are probably present in many other areas of California as well.

How to Find Brown Anoles: You can find Brown Anoles in numerous residential areas in Orange County and nearby areas. Simply go for a walk on a warm day through neighborhoods or in regional parks surrounded by residences, and watch for small brown lizards perched on branches, rocks, or running across yards and sidewalks. Like Green Anoles, Brown Anoles can be great entertainment, and I suggest watching them scarf down bugs, chase each other around, and "talk" to one another with their dewlaps and head-bobs. If you decide to try to catch one, be careful, as they tend to occur in yards, and you must respect people's privacy.

A male Jackson's Chameleon from California. *Photograph by Jeff Lemm.*

JACKSON'S CHAMELEON

TRIOCEROS JACKSONII

FAMILY CHAMAELEONIDAE

Chameleons in California? Yes, indeed! Throughout the country, there are numerous populations of Jackson's Chameleons that have been introduced, accidentally or on purpose, including in California. Native to high-elevation forests in eastern Africa, the Jackson's Chameleon thrives in tropical habitats where it was introduced, including in Florida and Hawai'i , along with several beachside areas in California that feature similar weather to that of its native range in Africa. One of these populations happens to be in Morro Bay, just down the highway from my university. Apparently, this population was founded in the 1980s when several chameleons escaped from a cage in someone's home. Long-time residents of the town say that chameleons soon spread throughout the neighborhood and, for a couple of decades, could be found in people's yards regularly. In 2008, I sent letters to residents in the neighborhood asking if they still saw chameleons. Some folks sent me photos of chameleons recently seen in their yards, while others noted that they suspected that cats had preyed on most of the lizards, as they hadn't seen many in recent years. One non-native species eating another! I got permission to search through a few yards for the lizards, and my students and I found one female lizard in a tree after many hours of searching. So, while the status of the population in Morro Bay is unknown, they definitely still occur there, or at least they did as of 2008. Although non-native species are never a good thing per se, it is unlikely that they do much harm in Morro Bay, given that they reside mainly in people's yards and have not spread into nearby natural areas over a period of forty years. The same cannot be said of introduced Jackson's Chameleons in Hawai'i , which are eating native insects and snails at an alarming rate. The conundrum of how to manage invasive species like the Jackson's Chameleon is a major one, but luckily California has been spared any major impacts from this species so far.

Appearance: Jackson's Chameleons are large lizards with long, curled tails and feet with opposing digits that allow them to grasp small branches on the trees in which they live. They are usually green, but they can change color rapidly and show off hues of gray, blue, yellow, and more. They have "bug-eyes" on the sides of their heads, and each eye can swivel around to enable the chameleon to see in all directions, including behind it. Male Jackson's Chameleons have three prominent horns on their faces that they use when butting heads (literally!) with other males over territories, and females do not.

Natural History: In California, Jackson's Chameleons are found mainly in residential neighborhoods in areas near their points of introduction. It is likely that chameleons have also escaped from captivity elsewhere in California but have thrived only in certain beachside areas due to the higher humidity and cool nights afforded by these foggy climates. They likely also rely on the water provided by irrigation in residential yards, making it unlikely that they will spread into wild areas. Although they tend to live in trees with small branches that they can easily grasp with their opposing digits and their tails, they can also be found walking across lawns or driveways. They eat by projecting their super-long, sticky tongues at an insect and hauling it back to their jaws. I could not find any information on the reproductive biology of Jackson's Chameleons in California. More generally, females give birth to live young that emerge in amniotic sacs and break free shortly after birth.

Range and Variations: Although little is known about their statuses, several populations of non-native Jackson's Chameleons have been reported in California: Morro Bay in San Luis Obispo County, Laguna Beach and nearby communities in Orange County, and Rancho Palos Verdes in Los Angeles County. There are also unsubstantiated rumors of a population in Balboa Park in San Diego County. If you see a Jackson's Chameleon in California, please take photos and post it to iNaturalist so that researchers can follow up.

This female Jackson's Chameleon was photographed in Hawai'i , but there is at least one population in California. *Photograph by Grayson Lloyd.*

How to Find Jackson's Chameleons: It is not very easy to find Jackson's Chameleons in California, mainly because the statuses of their populations are uncertain and because they occur in residential areas. If you decide to search for chameleons, be sure to avoid trespassing by getting permission from property owners. Search for chameleons in the canopies of trees that have the narrow branches that these lizards like, either in the daytime or using a strong flashlight beam at night.

A male Great Basin Collared Lizard keeps watch over his territory from an elevated perch. *Photograph by Chad Lane.*

GREAT BASIN COLLARED LIZARD

CROTAPHYTUS BICINCTORES

FAMILY CROTAPHYTIDAE

Collared lizards are what one might call a "badass lizard." It is irrefutable, really. These miniature dragons perch atop rocky outcrops, chase down smaller lizards, and smash their skulls with the incredible force of their thick, muscular jaws. When I was a student at UC Berkeley, I volunteered to help with a graduate student's research project studying how these animals eat. I captured some Great Basin Collared Lizards for his study, and I captured *a lot more* hatchling Western Fence Lizards in my parents' neighborhood for him to feed to the collared lizards in laboratory feeding trials. In the twenty-seven years since then, I have seen numerous species of collared lizards in the United States and Mexico, and their bold color patterns and amazing predatory behaviors make them some of the most beautiful and fascinating lizards ever. Although common in some areas, in many places in California you only see them now and then, making them a real treat to find in the wild. Sauroblivion: cured!

Appearance: Great Basin Collared Lizards get their name from the dark collar around their necks, which is always present. But watch out! Other species of lizards, including the Desert Spiny Lizard and the Banded Rock Lizard, also have dark collars. My students have made that mistake many times over the years. Unlike spiny lizards, which have spiky scales, collared lizards have smooth, granular scales. Their collar has two black bands sandwiching a white band. They are large lizards with proportionally large heads and thin necks, making them look like bobblehead toys. Great Basin Collared Lizards are dark brown with lots of white spots and sometimes little reddish-brown bands across their backs. They have muscular hind legs, and if the lizards are harassed by predators, their tails do not easily fall off. Males have larger heads than females do and more intense dark patches on their throats and crotches, and females sometimes have bright-orange spots along their sides.

Natural History: Great Basin Collared Lizards inhabit the eastern deserts of California and are most often found in areas with large rocky outcrops or small boulders. Their diet consists of numerous arthropods, like beetles and grasshoppers, plus small lizards including Common Side-blotched Lizards, Western Whiptails, and others. They sit on top of rocks and watch for bugs or lizards to scurry by, then run with great speed, often on their hind legs like little velociraptors, before catching the unfortunate prey item and smashing it with their strong jaw muscles. They also occasionally eat vegetation, including flowers. They mate in the spring, lay eggs in the early summer, and hatchlings appear in late summer. In some years, females can produce two clutches of eggs.

Range and Variations: Great Basin Collared Lizards occur throughout the Great Basin and Mojave Deserts, which in California includes the southeastern desert counties as well as eastern Lassen County. They also occur in much of Nevada plus southern Idaho, northern Arizona, and western Utah.

How to Find Great Basin Collared Lizards: Great Basin Collared Lizards are typically active when it is rather hot out. Search for them by hiking through rocky habitat and scanning the horizon for lizards sitting high on distant rocks. Often, they will run into burrows or under rocks before you get to them, so watching for them from afar is key. You can also drive slowly along roads through rocky habitat and watch from the window for the distinctive profile of these large-headed lizards sitting on rocks.

A colorful female Great Basin Collared Lizard.

Marisa Ishimatsu

BAJA CALIFORNIA COLLARED LIZARD

CROTAPHYTUS VESTIGIUM

You probably know that there is a greater diversity of lizards in Southern California than in Northern California. One obvious reason for this is climate: lizards are more common in hotter areas.

However, there is another reason that you might not know about. The Baja California peninsula in Mexico, south of San Diego, has an amazing history, and I am not talking about recent events, but rather deep biogeographic history over millions of years. What is now Baja California was a piece of coastal mainland Mexico until relatively recently (about 5 to 10 million years ago), when it broke off and inched northward as an island, eventually ramming into the North American continent, creating the San Jacinto Mountains and other nearby ranges in the process. Numerous species were thus isolated on an island for many generations, evolving independently, and then became connected with present-day California. The Baja California Collared Lizard is likely to be one such species. It occurs from central Baja California northward into a narrow, inland sliver in Southern California, where its range stops abruptly on the south side of the San Gorgonio Pass (the Interstate 10 corridor near Palm Springs). This pass appears to be the northern limit to the range of other cool lizard species, too, including the Granite Spiny Lizard and the Banded Rock Lizard, showing how ancient tectonic processes can impact where we find species today.

Appearance: The Baja California Collared Lizard looks very much like the Great Basin Collared Lizard (page 59). The two species differ in some subtle characteristics like the number of lip scales and the dimensions of the collar, but the best way to distinguish them is by location (see Range and Variations).

Natural History: Baja California Collared Lizards are found in dry, rocky habitats, where they can typically be spotted basking in the sun atop large rocky outcrops or small boulders spread around the desert floor. They are active during the day, often at very high temperatures, and they seek refuge under the rocks at night and when fleeing from predators. They eat numerous types of invertebrates as well as small lizards and occasionally vegetation. Mating takes place in the spring, females lay eggs in the summer, and these hatch in the late summer or early fall.

Range and Variations: Baja California Collared Lizards occur in California in a thin strip of desert habitat from the northern slope of the San Jacinto Mountains in Riverside County southward through eastern San Diego County and western Imperial County, then extend down through northern and central Baja California. They do not overlap with the range of the Great Basin Collared Lizard, which can be found right across the San Gorgonio Pass in the San Bernardino Mountains and extend throughout the Mojave and Great Basin Deserts.

How to Find Baja California Collared Lizards: As with the Great Basin Collared Lizard, the best way to find Baja California Collared Lizards is to hike or slowly drive through rocky habitats with little vegetation and scan for large lizards sitting atop rocks. They share habitat with several other large species of rock-dwelling lizards, for example the Granite Spiny Lizard (page 131), and with some experience you will be able to distinguish collared lizards from afar based on their proportionally large heads. When looking for collared lizards, binoculars can be useful for scanning rocky outcrops before you get too close.

BLUNT-NOSED LEOPARD LIZARD

GAMBELIA SILA

FAMILY CROTAPHYTIDAE

In no other Californian lizard species is the common name so inadequate a descriptor of the animal's qualities. Sure, the Blunt-nosed Leopard Lizard has a blunt nose, and it has a leopard-spot pattern, but let's face it—the name is boring. And this lizard is anything but boring. These large, speedy lizards zoom around the San Joaquin Desert like flightless dragons, terrorizing beetles and grasshoppers as they gorge on an entire year's worth of food in the few months of spring and summer in which bugs are available. Endemic to California, Blunt-nosed Leopard Lizards occur only in the San Joaquin Valley and adjacent slivers of habitat. Last time you drove through this southern part of the Central Valley, what did you see? Lots of farms, possibly huge clusters of oil-drilling rigs, but not much in the way of the original sparsely vegetated flatland that these lizards favor. As a result of this habitat conversion, Blunt-nosed Leopard Lizards have been extirpated from most of their historic range and only reside in tiny pockets of the San Joaquin Valley and the Carrizo Plain. They were designated as federally endangered in 1967, and researchers and land managers have continued to vigorously explore ways to protect their remaining populations ever since. My students and I have conducted field research on Blunt-nosed Leopard Lizards for several years, where we fitted tiny radio collars around their necks so we could follow them and study their behavior and physiology, and to collect data to understand how climate change may impact their already fragile populations. We found that the extreme temperatures of the San Joaquin Valley mean that the lizards can only be active above ground for a small period each day in the summer, and that the number of active hours will dwindle as temperatures in the desert continue to climb due to climate change. As if that isn't tough enough, the big lizards represent a tasty meal to many other desert inhabitants. One of my students texted me a photo from the field of a rattlesnake coiled up at the mouth of a burrow, saying that a particular lizard's radio collar signal was coming from within. I replied that the lizard was probably just somewhere deep in the burrow, behind the snake. When the student returned the next day, the snake was no longer sitting there, but the signal still beeped from within the burrow. It stayed that way for about two weeks, until the signal moved several meters away. When my student tracked it, she found the radio collar nestled in a large pile of snake dung on the desert floor, confirming that it had indeed become lunch for the rattlesnake. As you can see, it is tough being a Blunt-nosed Leopard Lizard. Despite these hurdles, these uniquely Californian lizards still cling to existence in various wildlife refuges and on public lands, challenging us to come up with innovative ways to protect them from extinction in the years to come.

Appearance: Blunt-nosed Leopard Lizards are one of California's larger lizards, with their bodies reaching up to about five inches in length, plus an even longer, narrow tail. As their name implies, their pattern is dominated by leopard spots. These dark spots are often separated by white bars, on a gray, tan, or pinkish back-ground. Seasonally, females show off gaudy bright-orange patches along their sides, and males develop salmon-colored patches along their necks and chests. They have large, well-muscled heads to facilitate chomping on thick-skinned arthropods, with a stubbier nose than the closely related and far more widespread Long-nosed Leopard Lizard (see page 70).

Natural History: Blunt-nosed Leopard Lizards live in desert and grassland habitat in the San Joaquin Desert of California, occu-pying flat or gently sloping areas, typically with hard, bare soils and few annual grasses. The lizards spend most of the year hiding from extreme temperatures deep inside the burrows of Giant Kangaroo Rats, another federally endangered species that lives only in central California. Adult lizards emerge from these burrows in March or April and are active above ground during the day until late summer, when they retreat into the burrows and stay there until the following spring. Mating occurs in the late spring, females

Female Blunt-nosed Leopard Lizards develop bright orange patches during the mating season. *Photograph by Max Roberts.*

lay eggs underground in the summer, and hatchlings appear in late summer and are active above ground through October. Blunt-nosed Leopard Lizards also rely on the shade provided by shrubs such as California jointfir (*Ephedra californica*) for shelter on hot days. Their diet consists primarily of arthropods like beetles and grasshoppers, which they eat mainly in the spring, along with the occasional small lizard. Predation by raptors and ravens, coach-whips and rattlesnakes, and getting run over by cars, may account for significant mortality in Blunt-nosed Leopard Lizards.

Range and Variations: Blunt-nosed Leopard Lizards are found only in isolated populations in the San Joaquin Valley and Carrizo Plain, from Merced County southward to extreme northeastern Santa Barbara County.

How to Find Blunt-nosed Leopard Lizards: Searching for Blunt-nosed Leopard Lizards is something that should be done with the utmost care. Due to their federally endangered status, it is unlawful to touch or capture Blunt-nosed Leopard Lizards without special permission from the US Fish and Wildlife Service, which is typically granted only to scientists and land managers. This doesn't mean that you are not allowed to search for them; it simply means that you must not interfere with them in any way, and you must instead admire them from a distance. I suggest driving along roads through likely Blunt-nosed Leopard Lizard habitat during the day in the late spring, when lizards are most active. They like to hang around on or near roads because these lack the invasive grasses that elsewhere block their views of insect prey and potential mates, so you might see one perched on the dirt berm of a road. For the love of lizards, be careful not to run them over when driving through their habitat, and never, ever drive off-road there. Bring binoculars so that you can observe them from a distance. If a lizard runs away from you, you have gotten too close to it. Blunt-nosed Leopard Lizards can also be found sitting under shrubs or inside the mouths of rodent burrows during the hot parts of the day, but if you walk around looking for them, exercise caution that you do not disturb them.

Long-nosed Leopard Lizard.

LONG-NOSED LEOPARD AND COPE'S LEOPARD LIZARDS

GAMBELIA WISLIZENII **AND** *G. COPEII*

FAMILY CROTAPHYTIDAE

Long-nosed Leopard Lizards (and the closely related Cope's Leopard Lizard) are by far some of the coolest lizards in California. Big, bold, beautiful, and smart—they check all the boxes. My favorite experience with a Long-nosed Leopard Lizard occurred when I took my herpetology students to the Mojave National Preserve on our annual desert field trip. One student called us all over to check out an amazing natural history event in progress. She stumbled across a large female Long-nosed Leopard Lizard holding a dead Desert Iguana in its jaws. While this is cool, it is not unusual, as leopard lizards eat other lizards on the regular. But this particular iguana was clearly too large for the leopard lizard to ingest. What happened next was incredible. With all of us watching (and filming), the leopard lizard dropped the iguana, grabbed its tail with her jaws, and with a practiced motion, tore the iguana's tail from its body, ate it, and walked off, leaving the rest of the body behind. This leopard lizard had killed an iguana that was too big for her to ingest, assessed this, and purposefully yanked off its tail to eat the smaller morsel. Now, Desert Iguana tails don't break off easily like in some skinks, geckos, or spiny lizards. It was as though the leopard lizard knew this, and she ripped it off with a complex rotating motion of her head that was evident in our filmed footage when we slowed it down. *She knew what she was doing.* This was a new behavior to science, and we dubbed it "saurocaudophagy" (*sauro* = lizard, *caudo* = tail, *phagy* = eating) in a natural history note that we published. There were a lot of excited students on that field trip, as well as a very excited professor. Notably, if we had jumped in and tried to catch the lizard, we would have missed this extraordinary behavior! Take the time to stand back and watch these incredible lizards in the wild—you never know what you will see.

Appearance: These large lizards get their names from their leopard-like spots, which are typically dark brown on a tan or gray background, interspersed with white crosshatches. Some individuals of Long-nosed Leopard Lizards (*Gambelia wislizenii*) may have a dark background with lighter spots and bands, and in Cope's Leopard Lizards (*Gambelia copeii*) there are typically paired dark spots along the spine that are separated by light bars. The spots extend onto the head in Long-nosed Leopard Lizards but not in Cope's Leopard Lizards. Females are slightly larger than males and often develop bright-orange patches along their sides during the mating season. These species have large, triangular heads, with longer snouts than Blunt-nosed Leopard Lizards.

Natural History: The Long-nosed Leopard and Cope's Leopard Lizards both inhabit dry flatlands with sparse vegetation that they can run between. They are active during the day, when they alternate between basking in the sun and hiding under shrubs from which they zoom out and capture passing arthropods and lizards. They dispatch these lizards by biting down repeatedly on their torsos, then ingest them headfirst. I once saw a Long-nosed Leopard Lizard walking about the desert with the tail of a Western Whiptail sticking out of its mouth, as it had easily swallowed the lizard's body but could not fit in the long tail until it had a little time to digest. Leopard lizards occasionally eat small quantities of vegetation, too. These lizards mate in spring, lay eggs in summer, and hatchlings appear later in the summer. They can sometimes lay multiple clutches of eggs in the same year.

Range and Variations: The Long-nosed Leopard Lizard is an extremely widespread species. In California, they occupy the southeastern deserts, then extend throughout much of the south-western United States into northern Mexico. The Cope's Leopard Lizard has a tiny range in extreme southern San Diego County and extends southward to occupy much of the Baja California Peninsula.

How to Find Long-nosed Leopard and Cope's Leopard Lizards: You should search for Long-nosed Leopard or Cope's Leopard Lizards on hot days by hiking or driving through excellent habitat, which includes desert flats with sparse vegetation like creosote bush

Cope's Leopard Lizard. *Photograph by Lee Grismer.*

and sagebrush. When hiking, you might flush leopard lizards out from under a bush, where the leopard pattern had allowed them to blend in with the shadows. They will usually run underneath a nearby bush, and if you approach slowly, you may be able to find them without scaring them again. Sometimes they flee into burrows instead. When driving, a passenger can scan the roadside berm or small rocks near the road for large basking lizards, which in this habitat are most often leopard lizards or Desert Iguanas.

Adult Switak's Banded Gecko.

Jeff Lemm

SWITAK'S BANDED GECKO

COLEONYX SWITAKI

I have never seen this lizard in California, though I have seen several individuals in Baja California, where most of its range occurs. The Switak's Banded Gecko, also known as the Barefoot Banded or the Peninsula Banded Gecko, is another example of a reptile species that occurs mainly on the Baja California peninsula and just barely gets into Southern California. As a result of its scarcity in the United States along with its rugged and remote habitat, rather little is known about its biology. It resembles the Western Banded Gecko (page 77) physically and in all likelihood ecologically, so I refer you to that account for more information. Also due to its scarcity in California, the Switak's Banded Gecko is listed as threatened in the state. Why do they also have the name *Barefoot*? I asked Dr. Lee Grismer, an expert in eublepharid geckos and Baja California reptiles, and he said that this adorable name likely comes from the fact that these geckos have a reduced number of scales on the bottoms of their toes.

Appearance: Switak's Banded Geckos are small, terrestrial (i.e., they don't climb on walls like geckos from the family Gekkonidae) lizards with prominent brown and pale spots in band-like patterns across the body and tail. Juveniles are bright yellow with black-and-white-banded tails, and males become rather yellow during the mating season. Their scales are smooth and granular but are punctuated by little bumpy tubercles. Their tails easily break off when harassed by predators and can regrow; the new skin typically lacks bands and instead may be spotted. Like Western Banded Geckos, Switak's Banded Geckos have eyelids and vertical pupils, and males have a pair of "spurs" at the base of their tail.

Natural History: In California, Switak's Banded Geckos occur in dry areas with plentiful large and medium-sized rocks and little vegetation. They hide inside rock crevices during the day and come out at night to feed on arthropods. Little is known about their natural history, including reproduction, but mating likely occurs in spring, and females lay one or two eggs at a time over an extended period in the summer.

Range and Variations: In California, the Switak's Banded gecko has an extremely limited distribution, occurring only in a tiny area of desert habitat in extreme eastern San Diego County and western Imperial County. From there, it extends southward as far as central Baja California.

How to Find Switak's Banded Geckos: This lizard is what herpetologists call "not a beginner lizard." Finding one requires a lot of effort, patience, and luck. It is difficult to drive around in most areas where these lizards occur because there are few roads, and Switak's Banded Geckos typically hide deep inside rock crevices during the day, so extended night hiking in rocky areas to look for active individuals is your best bet. Investing in an excellent, strong headlamp or flashlight is a good idea if you plan to engage in night hiking like this. Use your light to scan the rocky floor and look inside crevices. The best time of year to look for these geckos is May through July on warm and humid evenings.

Jeff Nordland

Hatchling Switak's Banded Geckos are bright yellow.

Spencer Riffle

WESTERN BANDED GECKO

COLEONYX VARIEGATUS

~~~~~~

FAMILY EUBLEPHARIDAE

This adorable little lizard will melt the heart of the most stalwart hater of wiggly creatures. Occasionally a reluctant student will enroll in my herpetology class, and the Western Banded Gecko converts them every time. Their eyelids make them look sleepy, and the shape of their jaws gives them the appearance of a wry grin. If you gently pick one up and shine your flashlight on its belly (and you will indeed have a flashlight when hunting geckos since they are nocturnal), you can often see its internal organs through its skin. On my annual herpetology class field trip to the Mojave Desert, we typically see a number of these geckos crawling across the roads at night, where they walk with their tails hoisted up above the ground. In headlight beams, they look like large scorpions trekking across the road. Geckos in this family are confused with scorpions throughout the world, with many people thinking they are venomous and can sting you with their tails. They definitely cannot, though they can in fact dispatch scorpions like nobody's business. Just like your pet dog "kills" its new toy, Western Banded Geckos grab scorpions and rapidly shake them before devouring them.

Appearance: Western Banded Geckos are little lizards with small legs, long toes, prominent eyelids, smooth granular scales, vertical pupils, and plump tails. Their toes do not grip walls like geckos in the family Gekkonidae; rather, they have pointy toes and walk on the ground. Their tails break off readily when touched by predators, and said tails then wiggle on the ground for several minutes to act as a distraction while the gecko gets away. Their color pattern is highly variable. Many individuals have irregular bands alternating between brown and yellow or pink, often with dark spots in the lighter bands. Although they are called banded geckos, in some areas they have spots only. Males can be distinguished from females because they have a pair of little "spurs" that extend sideways at the bases of their tails.

The Western Banded Gecko is one of California's only native nocturnal lizards. *Photograph by Jeff Martineau.*

Natural History: Western Banded Geckos live in deserts, chaparral, and arid woodlands, where they hide in burrows during the day and emerge at night to hunt. In the California deserts, these lizards are the only nocturnal lizards that are out and about on the desert flats, and as a result they enjoy a buffet of plentiful arthropods that only come out at night. However, they also are preyed upon by nocturnal snakes and other predators. Western Banded Geckos mate in the spring, and females lay large eggs one or two at a time throughout the summer. Amazingly, if you pick up a female gecko and look at her belly, you can easily see the eggs through her translucent skin.

Range and Variations: In California, one subspecies of Western Banded Gecko occurs in the southeastern deserts, and another approaches the coast in southern counties. They also occur in southern Nevada, Arizona, and extreme southwestern Utah, as well as Baja California and northern mainland Mexico.

How to Find Western Banded Geckos: If you read my earlier book *California Snakes and How to Find Them*, you know that night driving is a key method for searching for nocturnal snakes. Given that few California lizards are nocturnal, this method is not appropriate for most lizards. However, the exception is Western Banded Geckos. The best way to find them is to night-drive on desert roads with little traffic, especially those that transect rocky habitat. If you are new to night driving, drive slowly (about 20–25 mph) and watch for little lizards walking with their tails raised up like scorpions. You can also find Western Banded Geckos by flipping cover objects during the day, including palm fronds, boards or other objects at junk piles, and other debris. If you decide to capture a gecko, be sure to handle it gently, as its tail is liable to come off with just the slightest pressure, and its skin is fragile.

Jackson Shedd

ROUGH-TAILED BOWFOOT GECKO

CYRTOPODION SCABRUM

The Rough-tailed Bowfoot Gecko is one of the newest non-native lizards to hit the scene in California, first reported just five years ago. This lizard is native to the Middle East and northern Africa and is spreading throughout the United States by stowing away on commercial shipping containers and military transport planes. It first appeared in the United States in 1983 in a shipping yard in Galveston, Texas, where it became established. From there, it spread to Las Vegas and several cities in Arizona, and it appears that some lizards later hitched rides from Las Vegas to Fort Irwin and to Death Valley National Park in the Mojave Desert. Given that this lizard does very well in desert habitats and near human structures—both in its native range and in established non-native populations in Galveston, Las Vegas, and Yuma—it is highly likely that the Rough-tailed Bowfoot Gecko is becoming established and spreading throughout the California deserts as well.

Appearance: These small lizards have prominent, pointy, enlarged scales along the body and tail. They are tan with dark brown spots. Unlike many other geckos in the family Gekkonidae that have enlarged toe pads, the Rough-tailed Bowfoot Gecko has long, pointy, claw-like toes that it uses to climb around on surfaces including buildings. They have large, usually greenish eyes with narrow, vertical pupils and lack moveable eyelids.

Natural History: In their native range in the Middle East and northern Africa, the Rough-tailed Bowfoot Gecko lives in rocky desert habitat as well as in cities situated in these deserts. They are nocturnal geckos that climb on rocks and buildings, capturing insects attracted to the lights on buildings. Zilch is known about their reproduction in California, given that only a handful of specimens have been found so far. However, mating probably occurs in spring, with females laying several batches of one or two eggs throughout the summer.

Range and Variations: The Rough-tailed Bowfoot Gecko has been found in California in Death Valley National Park and Fort Irwin, and it is likely spreading to other areas of the Mojave Desert as I write this book. Its native range spans Middle Eastern countries, stretching from northeastern Africa to extreme western China.

How to Find Rough-tailed Bowfoot Geckos: Finding Rough-tailed Bowfoot Geckos in California might not be very easy, given that they have only recently been introduced to California and the status of any populations is uncertain. Also, given that the two known sightings occurred on a military installation and in a national park, permission may be needed to search for these lizards. If you find yourself in an area in which they might occur, you should look for them climbing on the outside walls of buildings on warm spring and summer nights.

HOUSE GECKOS

*HEMIDACTYLUS GARNOTII, H. MABOUIA, H. PLATYURUS,
AND H. TURCICUS*

FAMILY GEKKONIDAE

There are at least four established non-native gecko species from
the genus *Hemidactylus* in California. These lizards live and thrive
on buildings, and adults—and their eggs—can be unwittingly
transported around the world in everything from shipping containers to tourists' luggage, making them one of the most successful
invaders in all of lizarddom. The Mediterranean House Gecko is
widespread in the United States and has been around for a long
time, while the other species are more recent introductions and
occupy limited areas. I went to graduate school in Tempe, Arizona,
a place where the Mediterranean House Gecko has been established in urban areas for decades. There was a window above
our television, and we loved watching what we called "lizard TV"
along with our favorite shows. A group of geckos lived on the
outside of that window, and you could watch them pouncing on
flying insects, intimidating invading lizards with territorial displays,
and mating. They even had a little latrine, an area in the bottom
left corner where they all went to poop, keeping the rest of their
window-home nice and clean. I am grouping these four species of
lizards into a single account here based on their shared genus and
similar habits, even though they all come from different regions of
the Eastern hemisphere. I will highlight differences among species
in relevant sections below, although it's important to note that
these species are rather difficult to distinguish from one another,
as well as from lizards of the genus *Tarentola* (page 153), when you
are observing them from a distance. Those of you wishing to learn

Clockwise from top left: Indo-Pacific House Gecko in its native range in Myanmar, Tropical House Gecko, Mediterranean House Gecko, Flat-tailed Gecko in its native range in Indonesia.

the intricacies of all these species should invest in a detailed field guide on Californian reptiles or refer to CaliforniaHerps.com.

Appearance: All these species are small to medium-sized lizards with enlarged toe pads that they use to climb on buildings and ceilings, and eyes with vertical pupils and no moveable eyelids. Mediterranean House Geckos (*Hemidactylus turcicus*) are small lizards and are either pale pink or light brown with dark blotches, banded tails, and bumps all over their skin. Tropical House Geckos (*Hemidactylus mabouia*) are larger than the other species and are light in color with darker V-shaped blotches along their back, banded tails, and less-prominent skin bumps. Indo-Pacific House Geckos (*Hemidactylus garnotii*) are also small, are brown or gray with a mottled pattern and a yellow belly, with a fringe of sharp scales along each side of the tail, and otherwise have smooth skin. Flat-tailed Geckos (*Hemidactylus platyurus*) may be solid-colored or patterned and can be distinguished by their flattened tails with loose skin on the sides.

Natural History: Each of these species is primarily nocturnal and in California live mainly on the outside of buildings, especially under lights, where they feed on the insects attracted to the lighting. During the daytime they hide under roof shingles, in cracks, inside buildings, under vegetation, or any other cover objects they can find. They all eat invertebrates, and the larger Tropical House Gecko also sometimes eats hatchling lizards. For Mediterranean House Geckos, Tropical House Geckos, and Flat-tailed Geckos, males and females mate in spring and summer and possibly other times of year, then female lizards lay one or two eggs at a time often in communal nests. The Indo-Pacific House Gecko, on the other hand, is an example of an all-female parthenogenetic species that produces eggs on their own. This means that just one individual Indo-Pacific House Gecko can establish an entire population, partially explaining their successful invasions in California and elsewhere. Notably, even a single female with eggs, of any of these species, could also start a new population.

Range and Variations: Mediterranean House Geckos occur throughout much of California and are rapidly spreading to other areas of the state as well as much of the rest of North America. They arrived in the United States over a hundred years ago from their native range of the Middle East, northern Africa, and southern Europe, and were first documented in California in 1988. In contrast, other gecko species have arrived in California much more recently and still appear to be confined to small areas. Tropical House Geckos are native to Africa and have become established in Florida, some Caribbean islands, and several countries in South and Central America. They likely arrived in California via one of these populations, and they occur in one commercial area in Orange County and several residential areas in San Diego County. Indo-Pacific House Geckos are native to Southeast Asia and some Pacific islands and have established populations in the southeastern United States and Hawai'i . Flat-tailed Geckos are also native to Asia and are introduced in Florida. Each of these species has been found in housing developments in Los Angeles, Orange, and San Diego Counties. Notably, several other individuals of Tropical House Geckos and Indo-Pacific House Geckos have been found in other parts of California, but these do not appear to be established populations . . . yet. An additional species of gecko, the Fan-footed Gecko (*Ptyodactylus hasselquistii*) has also been introduced to California, but at the time of writing it has not spread beyond the industrial building complex in Orange County where it was spotted.

How to Find House Geckos: You can search for these lizards by looking for them hanging around on buildings under outdoor lights on warm nights. Bring a flashlight to help you see the lizards in detail. When lizard hunting in urban areas, it is extremely important that you exercise caution and respect people's privacy and private property. You probably wouldn't like it if a bunch of people carrying cameras and strange poles were shining flashlights around outside your bathroom window, would you?

Jeff Martineau

GILA MONSTER
HELODERMA SUSPECTUM

FAMILY HELODERMATIDAE

Ahhhh, the king! The Gila Monster (pronounced "hee-la") is the biggest, the most strikingly patterned, and the only venomous lizard species in California—indeed one of the only venomous lizard species in the world. In graduate school in Arizona, I got to do a side project radio-tracking Gila Monsters. These incredible beasts trundle around the desert, raiding nests for eggs and then sleeping off their meals inside burrows. It was such a privilege to be able to peer into their private lives for a couple of years. Arizona is the core area in this species' range, and they just barely cross over into extreme eastern California. So barely, in fact, that some people question whether viable populations actually exist in the state. There have been about thirty sightings of Gila Monsters in California in the past century and a half, and I only know of a couple of sightings in the state in the past two decades. Because of their rarity, and perhaps also because of their impressive size and the fact that they are venomous, Gila Monsters are high on everyone's list of lizard species to find in the wild.

Appearance: Gila Monsters are very large and stout lizards, the length of your forearm and about as big around. They have round, nubby scales that are reinforced with bone; these scales look like bubble wrap or little beads. Gila Monsters have a readily recognizable pattern of four mottled black bands that alternate with pinkish orange or light yellow, and their fat tails have solid bands.

Natural History: Gila Monsters live in remote, rocky mountain ranges along the eastern edge of California. They can be active day or night, and they hide in burrows or among rocks when temperatures are extreme. As nest raiders, Gila Monsters actively forage for eggs and babies of animals including birds, rodents, and tortoises. Scientists aren't sure if their venom plays a role in subduing prey, given that they eat animals that can't really defend themselves anyway. Their venom, which comes from glands in the lower jaw and slowly seeps into unfortunate victims as the monster clamps down, may be a deterrent against predators. Gila Monsters mate in the spring and lay eggs in the summer. Interestingly, the hatchlings don't appear until the following spring, suggesting that the eggs don't hatch for a long time or that the hatchlings stay underground through the winter.

Range and Variations: Gila Monsters range in extreme southern Nevada and Utah, through much of Arizona and extreme southeastern New Mexico, into northern Mexico. Most Californian specimens have been found in San Bernardino County, with a small number of individuals verified from Inyo, Riverside, and Imperial Counties.

How to Find Gila Monsters: Finding Gila Monsters in California is very difficult. It is the perfect storm of a rare animal: they spend a lot of time hiding underground, they live in remote, mountainous areas of the state with few roads to access lizard-hunting spots, and their populations are not particularly dense at the edge of their habitat. If you can find good roads, driving is a great tactic to look for Gila Monsters because you cover a lot of ground and might get lucky by seeing one out and about during mild weather. Many of the roads in their habitat are four-wheel drive only, so be careful. Another tactic is to hike in appropriate habitat and peer into burrows and rock crevices with flashlights or with sunlight reflected from your cell phone or a hand mirror. Note that handling Gila Monsters is prohibited, as they are designated as a species of special concern in California. If you do encounter one in California, be sure to get excellent photos and to record the exact locality to post to iNaturalist, as scientists will most certainly want to hear from you about such a finding. If you really want to see a Gila Monster, I suggest visiting the Sonoran Desert, for example the outskirts of Tucson, Arizona, and driving around remote roads or hiking on summer nights to look for these beautiful lizards, where they are still not common but are far more plentiful than in California.

DESERT IGUANA

DIPSOSAURUS DORSALIS

FAMILY IGUANIDAE

Desert Iguanas are very satisfying lizards. What do I mean by this? They are plentiful, they are large, they are relatively easy to find and watch, and they are adorable, especially when their mouths are stained yellow with flower goo in the springtime. Desert Iguanas like to be active when it is really hot out, so even on those days when it's over 100 degrees in the Mojave Desert, you are likely to see a few iguanas zooming around. When I was in college, I spent several days critter hunting at the Pisgah Lava Flow in San Bernardino County and got lucky with unseasonably warm weather at the end of March. Each day we woke up in our tents, went out to wander the lava rocks and collected data on all

Zeev Nitzan Ginsburg

the lizard species we could find. First out were the little Common Side-blotched Lizards, then the Western Whiptails, and when it warmed up a bit more we started seeing Common Chuckwallas, Great Basin Collared Lizards, and the occasional Desert Horned Lizard. In the middle of the day, when other lizards hid from the sun, out came the Desert Iguanas! A pair of them lived near our campsite, and when we returned from our morning hikes for lunch they would zoom at warp speed to the refuge of a burrow under a rock, then slowly creep back out if we sat motionless while eating our sandwiches. We dubbed them our camp hosts! Since that day, I have seen countless Desert Iguanas, and they are still just as cute, fascinating, and hot-blooded as that first trip to the Mojave.

Appearance: Desert Iguanas are large lizards with snub noses, fat bellies, and long, thick tails. They are rust-colored with rows of little white spots on their backs. These colors form a banded pattern on top, and their undersides are pale.

Natural History: Desert Iguanas are found in (duh!) the desert, especially sandy areas with sparse vegetation. They can also be found in rocky areas, as long as there is plenty of sand surrounding the rocks, and they are often found basking on small rocks. They are active during the heat of the day at temperatures that cause many other lizards to seek shelter. Desert Iguanas primarily eat vegetation, including flowers and leaves, especially those of the creosote bush, a plant with which they are commonly associated. When they have the opportunity, they will also eat insects and even dead animals like roadkill. Desert Iguanas mate in the spring, lay eggs in the summer, and hatchlings appear in late summer and early fall.

Range and Variations: Desert Iguanas occur in the southeastern deserts of California from Inyo and Kern Counties southward into Baja California and northern mainland Mexico, plus southern Nevada, extreme southwestern Utah, and western Arizona.

How to Find Desert Iguanas: Desert Iguanas are relatively easy to find. Walk around on warm or hot days in sandy areas in the desert, and chances are you will come across a Desert Iguana before too long. Choose areas with vegetation stippling the landscape, like creosote bush. The presence of rocks is also a positive, though not required. If you surprise a Desert Iguana, it will likely run away into a burrow or into the shade of a bush where they then freeze and stand still. In this way, their behavior is like that of Long-nosed Leopard Lizards. It is very common for me to be walking around the desert, flush a large lizard that flees into the shade of a creosote bush, and only after I slowly approach and look for it can I determine which of those two species it is.

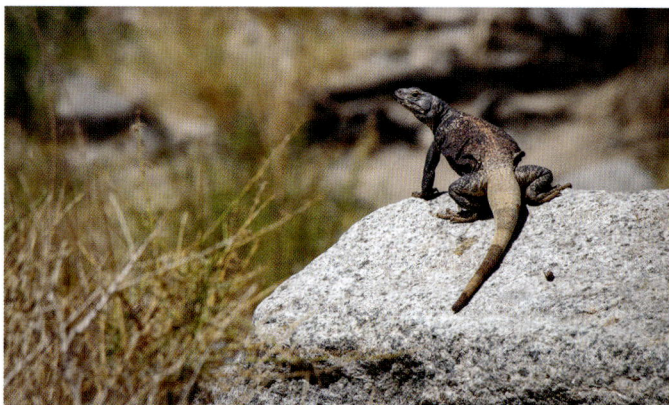

Zeev Nitzan Ginsburg

COMMON CHUCKWALLA

SAUROMALUS ATER

FAMILY IGUANIDAE

I have taught herpetology for over fifteen years, to hundreds of Cal Poly students, and I can confidently say that Common Chuckwallas (often just called Chuckwallas) are without a doubt the lizard species most loved by students. Heck, they are probably also my most loved lizard, and I am only being coy about giving them the trophy because I find it difficult to commit when it comes to lizards. Chuckwallas are so much fun! They are huge lizards (second only in size to Gila Monsters), and their enormous bellies make them practically as wide as they are long. Their bulging guts are supersized to house their extra-long intestines, which contain the multitude of symbiotic microorganisms that help Chuckwallas digest the plant material they eat. Chuckwallas are strictly a rock-dwelling species, and when they see you coming, they dive into crevices barely big enough to fit their fat bellies. If a predator (or a lizard enthusiast) tries to pull them out, they pump air into their lungs and puff themselves up, wedging themselves with great effect—it is often impossible to get them out! When my class visits the Mojave National Preserve, we have permits to catch and release Chuckwallas and other lizards so the students get hands-on experience with reptiles. Despite how difficult it can sometimes be to get our hands on a Chuckwalla, when we do, these photogenic critters are top features in my students' Instagram selfies.

Appearance: Chuckwallas are fat and flat, in other words wide but not super thick, allowing them to wedge inside those rock crevices. Their color varies quite a bit, often matching the color of the rocks on which they live. They are commonly gray, sometimes with white, yellow, or orange patches; and in areas with lava they are almost black and blend right in. Young Chuckwallas have banded tails, and the banding fades as they age, especially in males. Their scales are tiny and granular, like sandpaper.

Natural History: Chuckwallas live in desert habitats with complex rocky outcrops that have enough of the slit-like crevices in which they sleep and hide and sufficient vegetation to support their dietary needs. They are active only during the day, and at night they hide in their rock crevice refugia. They eat the flowers, fruits, and leaves of various desert plants, which they supplement with the occasional insect. Like most lizards, they mate in the spring, lay eggs in the summer, and hatchlings emerge in the early fall.

Range and Variations: Chuckwallas inhabit the southeastern deserts from southern Mono County southward to Imperial and inland San Diego Counties. Outside of the state, they range most of the way down the Baja California peninsula as well as

Juvenile Common Chuckwallas have strongly banded tails.
Photograph by Zeev Nitzan Ginsburg.

into northern mainland Mexico, plus western Arizona, southern Nevada, and extreme southwestern Utah.

How to Find Common Chuckwallas: Chuckwallas are as predictable and solid as the rocks on which they perch. Simply look in desert areas with rocky outcrops and plenty of creosote bush, and hike around during the late morning and midday to look for Chuckwallas basking on the rocks. Binoculars can be helpful because these lizards are tetchy beasts that will sometimes dive into their crevices before you even see them. If you spot a Chuckwalla basking, you need to approach very slowly to avoid spooking it. Another way to find Chuckwallas is to peek inside crevices when they are likely to be hiding there, including at night or at midday when temperatures are too extreme for them to be active.

ITALIAN WALL LIZARD

PODARCIS SICULUS

FAMILY LACERTIDAE

The year is 1994. A tourist from Los Angeles goes to Taormina, Sicily, and falls in love with the beautiful, jewel-hued lizards that crawl around on the rocky walls straddling the island. They decide to catch a few, smuggle them home in their luggage, and let them go in their suburban Los Angeles yard so that they can enjoy these beauties back home. Well, their plan worked. That is how the Italian Wall Lizard became established in San Pedro in Los Angeles County, where they have flourished ever since. It's a terrible idea to release non-native wildlife because they can outcompete or introduce diseases to native wildlife. I recently saw disturbing images online of veterinarians surgically removing huge worms from the lungs of a Florida Coral Snake; these worms arrived in the United States courtesy of an invasive species, likely the Burmese Python. It is not yet known whether the Italian Wall Lizard and other introduced lizard species have spread any parasites in California, but there is evidence that they could be excluding or outcompeting our native lizards, including Western Fence Lizards. Meanwhile, Italian Wall Lizards have since popped up in a couple of other neighborhoods in Southern California, and they are likely to keep appearing in new places, so please keep your eyes out for these lizards. Reports from community scientists like you are critical to learning about the spread of introduced lizards, as it is impossible for scientists to monitor all the neighborhoods in Southern California.

Appearance: Italian Wall Lizards are medium in size, with pointy snouts, beefy limbs, and long tails. Their color varies dramatically: they can be emerald green, brown, and yellow, sometimes solid-colored, and sometimes the colors form checkered stripes.

Natural History: In California, non-native Italian Wall Lizards live in suburban neighborhoods where they can be found basking during the day on rocks, walls, buildings, and other man-made structures. At night they hide in crevices or under rocks, shingles, or other objects found in people's yards. They eat mainly insects and small quantities of vegetation, and typically hunt their prey on the ground before returning to their perches. These lizards tend to dive into rock crevices or thick shrubs to escape from predators (or humans), but in California they are often fairly tolerant of people. Little is known about their reproduction in California, but it is likely similar to how things go in Italy given that both are

Mediterranean climates. Mating takes place in spring, and females lay multiple clutches of eggs one after another throughout the summer. Hatchlings can therefore appear throughout summer and fall.

Range and Variations: Although Italian Wall Lizards were first introduced to California in 1994, scientists were not aware of the San Pedro population until 2010. DNA analysis confirmed what the resident who had released them said about catching them in Sicily. In 2016, another population of Italian Wall Lizards was discovered in a neighborhood in San Marcos in San Diego County, but these lizards were an independent introduction because their DNA showed them to be from northern mainland Italy, which is a different subspecies than their Sicilian cousins. Apparently, the folks who purposely released Italian Wall Lizards into California neighborhoods were not alone in their decision to do so, because these lizards have also become established in multiple other cities across the United States and Canada. Italian Wall Lizards are native to—you guessed it—Italy, as well as parts of several neighboring countries from France eastward to Turkey.

How to Find Italian Wall Lizards: If you want to see an Italian Wall Lizard in California, I suggest walking around neighborhoods in the San Pedro neighborhood of Los Angeles on warm days when the lizards are out basking. As always when looking for wildlife in populated areas, be careful to follow the golden rule of urban creature hunting: behave as you would want visitors to your own neighborhood to behave. Be conscious of people's privacy and don't trespass.

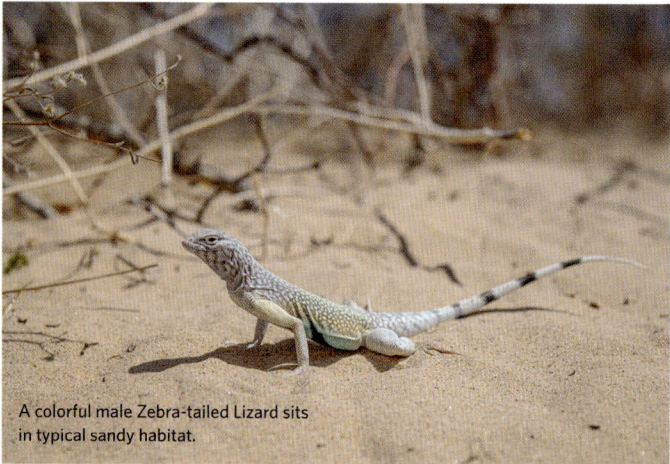

A colorful male Zebra-tailed Lizard sits in typical sandy habitat.

Marisa Ishimatsu

ZEBRA-TAILED LIZARD

CALLISAURUS DRACONOIDES

FAMILY PHRYNOSOMATIDAE

"This thing looks like it came from outer space." This statement, or something similar, has been uttered by numerous herpetology students over the years. Zebra-tailed Lizards are goofy-looking critters that charm everyone fortunate enough to see them up close and personal. Gangly bodies, bobbleheads with a snout like a little shovel equipped for digging sand, and quirky black and white tails that they wave at approaching humans, add up to make a darling beast. If that isn't enough, the vivid colors they develop in the spring put the cherry on top of this delicious lizard. Can you tell that I adore these lizards? They are very common in appropriate habitat in the desert, so go see for yourself—I dare you not to fall in love with our very own little Californian alien, the Zebra-tailed Lizard.

Appearance: Zebra-tailed Lizards are medium in size, with a round head that extends into a blunt, shovel-shaped snout; long, thin body and legs; plus a thin tail that they curl up like a scorpion to show off the black and white bands that are especially prominent on its underside. Their scales are granular and soft to the touch. They are a pale gray or tan, with tiny white spots stippling their backs. During mating season in the spring, males develop wildly showy colors on the sides of their bellies, specifically bright blue-green patches with two thick, diagonal black bars, all surrounded by light orange and yellow that extends onto the sides of their back like watercolor paint.

Natural History: Zebra-tailed Lizards inhabit sandy areas in the desert, especially areas with creosote bushes or other shrubs that are spread out on the landscape. They bask during the day on the hot sand, and when disturbed by a person they sprint away, sometimes to the cover of a shrub, sometimes to another exposed patch of sand, but often into the distance never to be found again. Zebra-tailed Lizards eat insects and arachnids like spiders, and very occasionally eat small lizards and desert plants. They mate in the spring (males are bright in color in the spring to attract mates), lay eggs in the summer, and hatchlings appear in the summer and early fall.

Range and Variations: Zebra-tailed Lizards live in the southeastern deserts from Mono County southward. They also inhabit much of Nevada, Arizona, southwestern Utah, and extend into northern Mexico and most of the Baja California peninsula.

How to Find Zebra-tailed Lizards: Zebra-tailed Lizards are very common in many sandy habitats in the desert. On warm days in the spring and summer, hike around in sandy washes or at the edges of sand dunes where vegetation begins. Zebra-tailed Lizards are typically active in late morning and midday, basking in the sun, though they will hide if it is extremely hot. If you're feeling a bit hot but not miserably so, then the temperatures are perfect for Zebra-tails. Keep your eyes trained on the sand in the distance as you walk, as the lizard will usually see you before you see it, and it will run off. If you watch where it runs to, you can slowly approach the spot and often find it sitting there.

A female Zebra-tailed Lizard lifts her toes off the ground to avoid gaining heat from it.

Max Roberts

Chad Lane

BANDED ROCK LIZARD

PETROSAURUS MEARNSI

FAMILY PHRYNOSOMATIDAE

Banded Rock Lizards, also known as Mearns's Rock Lizards, are quintessential rock-dwellers. After all, the name *Petrosaurus* means "rock lizard." Watching them run around on rocks is truly impressive. They are agile climbers, can run upside down on rock overhangs, and can jump from rock to rock with aplomb. Their bodies are flattened, and their legs extend out to the sides, as though someone set a brick on top of a lizard and compressed it top to bottom. This allows them to sidle into narrow rock crevices. I do what I consider to be a rather good imitation of this behavior for my students in class, by bending my arms and legs and spreading them out wide, then crab-shuffling out the door. Like many lizards you have read about in this book, Banded Rock Lizards occur mainly in the Baja California peninsula and just narrowly

Banded Rock Lizards are often found in groups. *Photograph by Max Roberts.*

extend into California. At the northernmost part of their range is a great spot where you can observe them along with another beautiful lizard, the Granite Spiny Lizard. Go to the Palm Springs Aerial Tramway and spend some time looking for lizards on the rocks around the tramway entrance. It's also very fun to take the tram up to the top of the San Jacinto Mountains; you zoom up 4,000 feet in elevation in just ten minutes, traveling from the desert to a forest dominated by fir and pine trees (and an entirely different community of reptiles).

Appearance: Banded Rock Lizards are medium-sized, flat lizards with narrow heads and rounded snouts; wide, pale bellies; and long tails. They are gray or brown on top, covered in blueish-black and white spots that usually give them a mottled banded pattern. Their tails are strongly banded, and they have a conspicuous black collar around the backs of their necks. Females get an orange hue on their heads during mating season.

Natural History: Banded Rock Lizards are 100 percent tied to rocks, especially large boulders in outcrops with plenty of narrow crevices for them to slip into. They bask in the sun on these rocks during the daytime, retreating into crevices in the middle of the day on exceptionally hot days. Banded Rock Lizards mainly eat insects and other arthropods, plus a certain amount of desert flowers and even the occasional small lizard. They mate in the spring, females lay eggs in the summer, and hatchlings appear in late summer and early fall.

Range and Variations: Banded Rock Lizards have a very small geographic range, existing only in northern Baja California and a narrow sliver of habitat in San Diego and Riverside Counties.

How to Find Banded Rock Lizards: Given that Banded Rock Lizards have a very restricted range in California, be sure that you go to the right place to find them. Much of their habitat lies in remote areas that are difficult to access, including canyons with plenty of vegetation, shade, and of course, large boulders. Hike around looking for lizards on large rocks and use binoculars to confirm the species, as several other lizard species live on rocks in the same areas as the Banded Rock Lizard.

Spencer Riffle

COAST HORNED LIZARD

PHRYNOSOMA BLAINVILLII

FAMILY PHRYNOSOMATIDAE

The Coast Horned Lizard, also known as the Blainville's Horned Lizard, is the subject of a sad and unfortunate story that is still being written in California. The two professors who taught herpetology at Cal Poly before me left records indicating that this lizard species was plentiful up and down coastal and central California. In recent decades, however, populations of Coast Horned Lizards have plummeted. Looking at most range maps won't give you the true story, as the lizards are totally gone from huge parts of their previous range and are isolated to certain pockets of habitat only.

They are not nearly as dense as they used to be in most places in their range. Why is this? One reason is that they generally don't thrive in areas around humans, and massive development continues to take place throughout their range in California. Another major reason is the invasion of North America by Argentine ants. These ants arrived in the American South at the end of the nineteenth century as stowaways on cargo ships from South America, and then marched overland to spread throughout much of the country. In California, they live mainly in coastal areas, so their range overlaps rather well with that of the Coast Horned Lizard. The problem is that the Argentine ants outcompeted the native ant species on which the horned lizards feed, and they cannot seem to switch over to eating these invasive ants. As a result, populations of Coast Horned Lizards have slowly gone extinct throughout the state, making them now a California species of special concern, possibly headed toward threatened status. When you do find one of these lizards, they sure are a treat. You may have heard them called "horny toads," a nod to how their fat bodies and snub noses give them a frog-like appearance. The name *Phrynosoma* means "body of a toad," after all.

Appearance: It is difficult to mistake a Coast Horned Lizard for anything else. All horned lizards are flat, with extremely wide bellies, snub noses, short legs and tails, a prominent crown of horns around the head, and many small spikes sticking out of their backs. The Coast Horned Lizard is distinguished by having two rows of fringed scales along each side of the belly, and the center pair of horns on the head is often slightly enlarged. Their color is highly variable and can include spots of gray, brown, red, and orange.

Natural History: Coast Horned Lizards are found in open areas with sandy soils in beaches, grasslands, chaparral, and woodland. They are most common at low elevations but in some areas can extend up to 8,000 feet in elevation. They are active during the day. Like other horned lizard species, they eat harvester ants and other invertebrates, and can often be found hanging around anthills. Mating occurs in spring, females lay one or two clutches of eggs in the summer, and hatchlings appear in summer and early fall.

Range and Variations: The geographic range of Coast Horned Lizards extends from the south Bay Area in coastal counties down just barely into Baja California, where they then transition into related species of horned lizards that extend farther down that peninsula. They also occupy California's Central Valley and foothills from Butte County southward, though much of that habitat has been converted to agriculture, making lizard populations spotty.

How to Find Coast Horned Lizards: Searching for Coast Horned Lizards can be maddening because plenty of spots that look like they should be good habitat are devoid of lizards, presumably due to the impacts of Argentine ants. You will need to just keep at it. Hike around during the day in sandy areas in the spring and watch for these stout little lizards. They are quicker than they look like they should be, yet they can't run very far. If you flush one, it will likely run a short distance to hide under a shrub. If you decide to catch one, handle it with care! When roughly handled, these lizards can literally shoot blood at you out of their eye sockets. Yep, you read that right. Plus, their blood is caustic from eating the formic acid in all those ants, so it is not going to feel very good when an attacking coyote or fox, or a handsy lizard hunter, gets a spray of blood in their eyes.

Francesca Heras

Spencer Riffle

PYGMY SHORT-HORNED LIZARD

PHRYNOSOMA DOUGLASII

FAMILY PHRYNOSOMATIDAE

The Pygmy Short-horned Lizard is a rare case of a lizard that is restricted to Northern rather than Southern California. As discussed in the introduction, the cooler weather in their range may be responsible for an adaptation that they share with just a handful of other California lizards: live bearing. By keeping the embryos inside her body instead of laying them in eggs, a pregnant female can bask in the sun and raise her body temperature significantly higher than that of the air around her. Then she gives birth to tiny lizardlets. Which brings me to another point: baby horned lizards are some of the most adorable creatures on the planet. (See the top photo on page 25 for evidence.) They are about the size of a bottle cap, their horns are even smaller than those of their parents, and they have a grumpy frown and side-eye that rival any animal you have seen. Hiking around in their habitat in August and September can sometimes yield the mother lode: small groups of teeny baby horned lizards scattering from your feet as you walk down a path in a sunny clearing in the forest.

Appearance: The name Pygmy Short-horned Lizard tells you two important things about their appearance: they are small and they have very short horns. Like other horned lizards, they have a wide, fat body with a short tail and snub nose, and their backs are covered in pointy scales. They have a single row of fringed scales along each side of the body. The color of these lizards varies tremendously and often matches the color of the substrate to help them blend in. A series of two dark spots outlined in white form rows down their backs.

Natural History: Pygmy Short-horned Lizards are found in open areas with sandy soils in woodlands and rocky areas. They eat ants and other invertebrates. Mating occurs when the lizards emerge from hibernation in the spring, and females give birth to live young in the late summer.

Range and Variations: In California, Pygmy Short-horned Lizards are found mainly in Siskiyou County, with possible populations in nearby counties. From there, the lizards extend into northern Nevada, Idaho, eastern Oregon and Washington, and just barely into southern British Columbia.

How to Find Pygmy Short-horned Lizards: Hike around in sandy meadows or clearings in forests in the mountains of Siskiyou County on warm summer days, watching for these little lizards basking on the ground or on small rocks in sunny areas. If they see you before you see them, they might run into the shade of sagebrush or other vegetation. You can slowly walk up and find the lizard sitting there, giving you its endearing scowl.

Chad Lane

FLAT-TAILED HORNED LIZARD

PHRYNOSOMA MCALLII

FAMILY PHRYNOSOMATIDAE

Flat-tailed Horned Lizards have the distinction of occupying the smallest geographic range of any horned lizard in California. Its entire range, which extends a bit into Arizona and northern Mexico, is not much larger. The areas in California and Arizona where the Flat-tailed Horned Lizard lives are characterized by high rates of development, making this lizard a candidate for government protection. It actually was federally protected in 1993, but protections were withdrawn in 2006. California decided not to give the lizard threatened status, opting to designate it as a

Marisa Ishimatsu

species of special concern. The reasons behind such decisions are complicated and often political in nature, involving a balancing act between protecting biodiversity and allowing people to use the land the species occupies for development or recreation. Flat-tailed Horned Lizards are impacted by all sorts of problems, including agriculture, off-road vehicle use, and invasive vegetation that mucks up the open sandy areas that they enjoy. The vegetation, like invasive mustard, is also a problem because the Flat-tailed Horned Lizard's favorite ant prey dislikes living in areas with thick vegetation. The Flat-tailed Horned Lizard is one of the most studied lizards in California despite its tiny range, as scientists try to figure out ways to help it persist in the face of extreme change.

Appearance: While the entire body of the Flat-tailed Horned Lizard is flattened, as with all horned lizards, its tail is particularly flat and elongate. These medium-sized lizards have a prominent crown of eight horns on the head, and the center two horns are long and sharp. Their backs are not as spiky as most other horned lizards'. They have an obvious dark stripe down the center of their backs, and several pairs of dark spots alongside the stripe. They are usually tan, with variations that match the color of the fine sand on which they live.

Natural History: Flat-tailed Horned Lizards are found in areas with gravel, compacted, or loose sand, including the edges of wind-swept sand dunes with sparse vegetation. Many of these areas are popular among off-road vehicle users in the desert, which is problematic because the vehicles can destroy habitat used by the lizards and can even run them over. Flat-tailed Horned Lizards are ant specialists, and occasionally eat other insects. These lizards mate in the spring, lay eggs in the early summer, and hatchlings arrive in the late summer.

Range and Variations: Flat-tailed Horned Lizards inhabit low elevation areas including the Coachella Valley of Riverside County, the Imperial Valley in Imperial County, and the vicinity of Anza-Borrego Desert State Park at the border of San Diego and Imperial Counties. They also extend into extreme southwestern Arizona near Yuma, and into northern Mexico.

How to Find Flat-tailed Horned Lizards: Find areas in Anza-Borrego or the Coachella or Imperial Valleys with sand dunes or sandy or gravelly soil interspersed with shrubs and walk around on warm days looking for these little lizards. Flat-tailed Horned Lizards often freeze when they see a predator or human and can go undetected because they blend into the sand. If they do run away, you'll be amazed at how quick they can be. Follow them in a hurry, because they are apt to burrow themselves into the sand to get away from you!

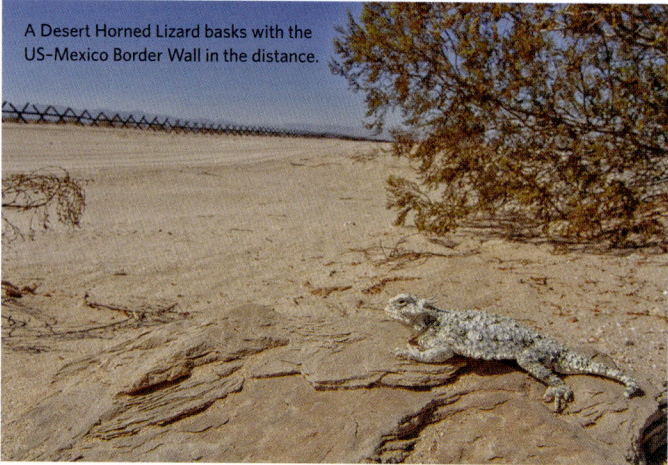
A Desert Horned Lizard basks with the US–Mexico Border Wall in the distance.

Marisa Ishimatsu

DESERT HORNED LIZARD

PHRYNOSOMA PLATYRHINOS

FAMILY PHRYNOSOMATIDAE

One of the most charismatic lizards in California's deserts is the Desert Horned Lizard. Colored just like desert rocks and sand and covered in poky little spines to help protect them from becoming the snacks of birds and other predators, these lizards draw wows and smiles from anyone lucky enough to encounter them. A Desert Horned Lizard was the very first species of horned lizard I saw in the wild. I was wandering around the Pisgah Lava Flow in San Bernardino County on a warm day, when a little chunk of lava on the ground near my feet scuttled into the shade of a creosote bush. I did a double take and saw that it was actually a horned lizard! It peered suspiciously at me as I approached and slowly squatted down next to it. To be honest, horned lizards always appear suspicious; that's just the way their faces are shaped. One of my students lovingly coined the term "resting brat face" to describe these adorable little lizards' visages. We usually find just one or two on any given field trip, and rarity breeds excitement. Also, people just love lizards with fat bellies. Between Desert Horned Lizards, Common Chuckwallas, and Desert Iguanas, the fat belly department is covered in California. Desert Horned Lizards are one of the most popular lizards among my students on our annual desert field trip. Clearly fond of coining terms, my students also came up with a name for their photos with horned lizards: Phrynoselfies!

Appearance: Desert Horned Lizards are medium-sized lizards that are flattened top to bottom and have a large, round belly. They have a crown of horns around their head, with the central two horns enlarged, plus numerous sharp, spiky scales sticking out of their backs. A single row of fringed scales sits along each edge of the belly. Their coloration is highly variable, and typically matches the substrate of their habitat. For example, the lizard that I saw on the lava flow was dark gray with brown and dark-red spots, while Desert Horned Lizards in sandy areas tend to be tan with white and brownish-orange spots.

Natural History: Desert Horned Lizards are typically associated with dunes, washes, or patches of sand among rocky outcrops. They bask in the sun during the day, lying in the sand or sitting on small rocks. Like other horned lizards, they specialize in eating ants, and will occasionally eat other invertebrates. Mating occurs in spring, females lay one or two clutches of eggs in the summer, and hatchlings appear in the late summer and early fall.

Range and Variations: Desert Horned Lizards are a quintessential desert lizard, spanning the Great Basin Desert, Mojave Desert, and part of the Sonoran Desert across multiple states including Oregon, Idaho, Nevada, Utah, Arizona, and California, into northern Mexico. In California, they are found in the southeastern deserts from Mono County south to Imperial County, and their range also extends slightly into the northern deserts in Modoc and Lassen Counties.

How to Find Desert Horned Lizards: Hike around in sandy areas of the desert on warm days, scanning the ground for lizards as you go. You will likely see multiple species in such habitat, and hopefully will encounter a Desert Horned Lizard or two! They will typically run a short distance when disturbed, making it easy to catch up to them and take a close look. If you want to catch them, you can usually just pick them up by hand—no need to use a lasso. Admire them, take a Phrynoselfie, and then let the lizard go. Resist the urge to collect these lizards to keep as pets. They do not thrive in captivity, in part because they insist on eating specific ant species only.

Zeev Nitzan Ginsburg

COMMON SAGEBRUSH LIZARD

SCELOPORUS GRACIOSUS

FAMILY PHRYNOSOMATIDAE

Common Sagebrush Lizards are very common lizards at higher elevations throughout California and much of the American West, and in some areas, they extend into lower elevations as well. These showy little lizards might be common but are nonetheless worthy of your attention. The males have super-colorful throats and bellies that they show off to one another when engaging in disputes over perches and females, and these behaviors are delightful to sit and watch while on a hike in the mountains. In summer of 2015, my research group conducted a two-week field study aimed at understanding how elevation affected thermoregulation in the Western Fence Lizard. We pored over maps of the

This female lizard in breeding coloration is from a Southern California population that is considered by some to be a separate species, the Southern Sagebrush Lizard (*Sceloporus vandenburgianus*).

Sierra Nevada to identify sites at different elevations from which to capture lizards. We started at a lower elevation site along the Kern River, where we spent a few days collecting data on lots of Western Fence Lizards, then went to another site about two thousand feet higher and repeated. We noticed that the lizards were larger at this higher site. Then we went to a third site, higher yet, and the Western Fence Lizards were much larger (huge, in fact!), plus loads of baby lizards were running around everywhere. At least that is what we thought until we captured one. Once the little lizard was in my hands, I laughed out loud as I realized my mistake. The small lizard was actually a Common Sagebrush Lizard, a species that only occurs at higher elevations in this part of the Sierra Nevada. At our fourth and final site, at about 9,000 feet, there were only Common Sagebrush Lizards present and no longer any Western Fence Lizards. It was fascinating how the Western Fence Lizards grew larger as elevation increased, so at the only site where the two species both occurred, they differed dramatically in size. This means that they probably eat different insects, use different perches and shelters, and otherwise use different resources and avoid competing with one another.

Appearance: Common Sagebrush Lizards are small lizards with spiky scales on their backs and tails. Their thighs have smooth, granular scales, which is one way to distinguish them from young individuals of the somewhat similar-looking Western Fence Lizard, which has spiky scales on their thighs as well. They are brown or gray with darker spots that often align to form a series of regular or irregular stripes that extend down their backs. Males have beautiful blue patches on their chins and bellies, and during mating season both males and females sometimes develop orange washes along their sides.

Natural History: Common Sagebrush Lizards certainly inhabit areas with sagebrush as the dominant plant, and they also occur in chaparral and in open areas in pine forests. They bask during the day on rocks, trees, and on the ground. They eat small insects and arachnids. They often have a short and compressed active season, especially those populations at high elevations, where snow covers the ground for much of the year. So, they begin mating as soon as they emerge from hibernation in the late spring, and breeding can continue through the summer. Females lay one or two clutches of eggs in the summer, and hatchlings appear in the late summer.

Range and Variations: Common Sagebrush Lizards have a complicated geographic distribution in California, with three subspecies occupying various mountain ranges and, in some areas, extending into lower elevations. Some people consider a southern subspecies to be its own species (Southern Sagebrush Lizard, *Sceloporus vandenburgianus*). Common Sagebrush Lizards occupy most of the Northern Coast Range mountains as well as the Sierra Nevada and the Great Basin Desert to the east, and in Southern California they are found at high elevations only in some of the mountains of the Peninsular and Transverse Ranges. Outside of California they have a large range, extending throughout much of the American West until their range ends in North Dakota, Nebraska, and Colorado.

How to Find Common Sagebrush Lizards: Common Sagebrush Lizards are very easy to find. As sun baskers, they are conspicuous, and they tend to be rather abundant in most parts of their range. Go for a hike in an area where these lizards occur and simply look for small lizards basking on any surface you can imagine. Depending on where you are, you'll need to be sure you can distinguish between adult Common Sagebrush Lizards and baby Western Fence Lizards. Lizards in the genus *Sceloporus* are the most lasso-able of all California lizards, so if you want to capture them to admire them up close, that is your best bet.

DESERT SPINY LIZARD

SCELOPORUS MAGISTER

FAMILY PHRYNOSOMATIDAE

When people think of the Mojave Desert, they often picture Joshua trees. So do I, but the ones in my mind's eye feature a pair of colorful Desert Spiny Lizards running up and down them. The first lizard species I studied as a brand-new assistant professor at Cal Poly in 2006 was the Desert Spiny Lizard. Every weekend in the spring, my students and I would pack up our camping gear and head out to a thick Joshua tree forest near Palmdale, where we drew blood samples from as many spiny lizards as we could get

Male (left) and female (right) Desert Spiny Lizards rest on a rock together.
Photograph by Zeev Nitzan Ginsburg.

our hands on. We were studying how the hormone testosterone is related to their growth patterns. A colleague had recommended this particular Joshua tree forest due to its dense population of lizards. Sure enough, every tree seemed to have a pair of lizards, one male and one female, who would run up to the top of their tree and stare down at us with disdain, daring us to try to catch them. This trip put our lizard lassoing skills to a serious test, as Desert Spiny Lizards can be tricky beasts, especially when it is hot out. Up the lasso would go, and the lizard would run to the other side. We'd move to the other side of the tree and try again, and the lizard would just run back. I swear I saw one roll its eyes at me. I consider myself to be an expert lizard lassoer these days, in no small part due to the experience I got chasing these tricksters around (and around, and around) Joshua trees back in the day.

Appearance: These are large lizards with heavily spiked scales all over their heads, backs, and legs, and a prominent, thick black collar. Their color varies, but can include backgrounds of yellow, beige, gray, or brown, often with dark and sometimes also light markings that form band-like patterns across their backs and tails.

Natural History: These lizards are widely distributed in arid habitats, where you can find them perching during the day on rocks, many kinds of plants (including Joshua trees), and on the ground near burrows where they take refuge. They are mainly insectivorous, but lizards this large are bound to also eat the occasional small lizard or even a rodent or bird. Mating is in spring and summer, females lay eggs in summer, and hatchlings arrive in late summer.

Range and Variations: In 2006, what was formerly one species was divided into two: the more northern Yellow-backed Spiny Lizard and the more southern Desert Spiny Lizard. Most herpetologists think that they should not have been split into two species, so I will consider them here as subspecies. The Desert Spiny Lizard subspecies occurs in the extreme southeastern part of the state, including Imperial, San Diego, Riverside, and southeastern San Bernardino Counties. The Yellow-backed Spiny Lizard occupies the eastern deserts from Mono County south to northern Riverside County, and its range extends in a spotty fashion westward to poke into San Benito, Fresno, San Luis Obispo, and Ventura Counties. Collectively, the species also extends into northern Baja California and mainland Mexico as well as into Nevada, Arizona, and tiny pieces of New Mexico, Utah, and Colorado.

How to Find Desert Spiny Lizards: In many parts of the desert, these lizards are not tremendously common, and you must be persistent to find them. The best way to find them is to hike around in their habitat during the late morning on warm days, scanning for them sitting on rocks or trees. However, in some parts of the desert, these lizards can be extremely common. Certain areas with isolated rocky outcrops or with thick Joshua tree forests can host very high densities of these lizards. For example, in the Mojave National Preserve you can choose any area with dense Joshua trees, and when you walk around on a warm spring day you will see Desert Spiny Lizards climbing around on almost every tree.

A Western Fence Lizard. *Photograph by Spencer Riffle.*

FENCE LIZARDS

SCELOPORUS BECKI AND *S. OCCIDENTALIS*

FAMILY PHRYNOSOMATIDAE

Western Fence Lizards are the most wide-ranging and common lizard species in California, and probably the most underrated. Any child with an interest in chasing creatures around in the outdoors will be familiar with these "blue-bellies" that grace so many rocks, trees, and fences in our state. Sometimes lizard enthusiasts

An Island Fence Lizard on Santa Cruz Island. *Photograph by Jackson Shedd.*

become bored with super-common species because they see them all the time. But Western Fence Lizards are exceedingly beautiful animals with a fascinating natural history to boot. When I took herpetology in college, my professor taught us how to catch these lizards with little lassos made from dental floss attached to the ends of sticks. The weekend after I learned this, I was home visiting my parents and I talked my father into letting me transform his fishing rod into a lizard pole. I went out that afternoon and caught lizard after lizard, just for fun, letting them go after admiring them. They have such beautiful color patterns, especially the males decked out in their flirtatious mating colors. Aside from Western Rattlesnakes, these lizards are the species my students and I have studied the most over the past twenty years. We have caught thousands of these lizards, studying everything from how their hormone levels influence the number of ticks that feed upon them to how climatic conditions influence how much water evaporates across their skin. Speaking of ticks, Western Fence Lizards are famous because when they are bitten by ticks that are infected with the bacteria that causes Lyme disease, the lizards' immune systems kill the bacteria, thereby preventing that tick from infecting another animal (including humans!) during its next bite. Due to their beauty and their fascinating physiology, every individual Western Fence Lizard is a joy to behold, with their spiky scales, the bewildering spectrum of blue on the belly, the tick "earrings" that adorn their heads each spring, and the hilarious push-ups they do when other lizards invade their spaces.

Appearance: Western Fence Lizards are small- to medium-sized lizards with spiky scales all over their bodies. They are usually gray or brown, with V-shaped marks or stripes of dark and light brown, yellow, or white on their backs. There can also be flecks of bright-green, blue, or other colors along the back, especially in males. Males have bright-blue patches on their bellies and sometimes their chins, and females often also have these patches, though they are usually fainter than in males. Given how widespread this species is, it is possible to find individuals with many other color and pattern variations as well.

Natural History: Western Fence Lizards are active during the day, when they bask in the sun on the ground, rocks, trees, man-made structures like fences, and other surfaces. They can be found in most any habitat in California, from the coastal beaches and chaparral, to all types of woodlands spanning from sea level to high elevations in the Sierra Nevada and Southern California mountains, to thousands of yards in suburban and urban areas throughout the state. Their diet consists mainly of insects and arachnids, though they also have the distinction of being known to occasionally cannibalize young Western Fence Lizards. They mate in the spring, a time coinciding with intensified colors and push-up displays by the males. Females lay one or more batches of eggs throughout the spring and summer, and hatchlings appear in the summer and fall.

Range and Variations: Western Fence Lizards (*Sceloporus occidentalis*) occur almost everywhere in California except for the arid lowland deserts of the southeastern parts of the state. Even in the deserts, these lizards occupy the tops of mountain ranges that form forest islands in the sky. Western Fence Lizards also occur northward into northern Washington, eastward into western Idaho and Utah, and southward into northern Mexico. The lizards on the Channel Islands have historically been considered a subspecies but recently were elevated to their own species, the Island Fence Lizard (*Sceloporus becki*).

How to Find Fence Lizards: The better prompt here would be how *not* to find a Western Fence Lizard. These lizards are *everywhere*. Walk around any natural area on a warm (but not too hot) day, and look for these lizards perched atop rocks, climbing up oak trees, basking on the ground and running for the shelter of a bush, or slinking along wood fences, and you will surely see many Western Fence Lizards. You can also sometimes flip cover objects at night or in cool weather and find these lizards slumbering underneath. Interestingly, these lizards are absent from many urban areas where there are too few trees and other perches for them to use, where pesticides kill their insect prey, and where outdoor cats hunt down any lizards intrepid enough to brave the city.

Max Roberts

GRANITE SPINY LIZARD

SCELOPORUS ORCUTTI

FAMILY PHRYNOSOMATIDAE

The Granite Spiny Lizard is California's most beautiful lizard. Sure, beauty is in the eye of the beholder, but if you get one of these lizards in your hands, you won't disagree with me. Large male Granite Spiny Lizards have shocking coloration. On top they are covered in various iridescent shades of purple, copper, green, and black. Underneath they have huge, bright-blue belly patches, sometimes extending literally from the snout to tail tip and from fingertips to toe tips. This species of lizard was the subject of my very first scientific study in college, to fulfill the capstone assignment of a famous class at UC Berkeley—Natural History of the Vertebrates. We were required to conduct our own field study, analyze the data, and submit a short scientific-style paper

Max Roberts

Granite Spiny Lizards shelter in crevices within granite outcrops.

reporting its results to the class instructors. I decided to go to the desert with a classmate and compare the flight initiation distance of several lizard species. I would approach each lizard and then when the lizard noticed me and dove for cover, I would freeze and measure my distance to the lizard. A low flight initiation distance meant that the lizard let me get close to it before it fled, which indicated it was less afraid of people. Well, guess which lizard species had a super-*high* flight initiation distance? Yep—Granite Spiny Lizards. These hyper-wary lizards often dove to hide in rock crevices when I was still forty feet away! So, this is a perfect species to spot and watch from a distance with binoculars. Just make sure you make a point of going on a trip to herp for this species so that you can see this most beautiful of Californian lizards.

Appearance: Granite Spiny Lizards are large-bodied, dark lizards with very spiky scales on their bodies, limbs, and especially their tails. Color is variable, but generally these lizards appear dark overall with a copper background marked with wavy dark bands and a black collar. Some males have bright iridescent green sides and purple backs, with purple occasionally covering the whole back. The entire underside can be a vivid blue.

Natural History: Granite Spiny Lizards are associated with rocks, especially large boulders in complex outcroppings with plenty of interspersed desert vegetation. They bask on these rocks during the day, with males occupying the tallest perches, and hide from predators and from the elements between and under rocks and in crevices. Granite Spiny Lizards eat mainly arthropods like spiders, beetles, flies, and grasshoppers, plus occasional flowers, and large individuals sometimes eat hatchling lizards. Mating takes place in spring, females lay eggs in late spring and summer, and hatchlings appear in summer and fall.

Range and Variations: Granite Spiny Lizards are a Baja California species that extends a short distance into California, mainly in San Diego and western Riverside Counties plus small parts of Orange, Imperial, and San Bernardino Counties. They are present at a wide span of elevations.

How to Find Granite Spiny Lizards: Granite Spiny Lizards are extremely conspicuous. As their name suggests, they frequent granite boulders, and the large, dark lizards are easily seen atop these light-colored rocks. They are sufficiently wary that getting close to them can be difficult, so I recommend observing them via binoculars. You can try to sneak up on them by walking very slowly and ducking behind vegetation as you approach. It's also easier to approach these lizards in the mornings when they first emerge than it is when they are hot in midday and the afternoon. Finally, in the evenings or on scorching days when it's too hot even for these baskers to be out, you can peer into crevices on granite boulders to look for adorable little faces peeking back out at you.

Mohave Fringe-toed Lizard. *Photograph by Marisa Ishimatsu.*

FRINGE-TOED LIZARDS

UMA INORNATA, U. NOTATA, **AND** *U. SCOPARIA*

FAMILY PHRYNOSOMATIDAE

The three species of fringe-toed lizards that occur in California share fascinating adaptations for living in habitats with fine, loose sand, including sand dunes and washes. First is the namesake fringe on the long toe of each hind foot, which helps them to run across the sand. Next, their nose looks a bit like an upturned shovel, which they use to burrow into the sand to get away from predators and to bed down for the night. All the openings on their head have adaptations to keep sand out when burrowing,

Colorado Desert Fringe-toed Lizard.

including valved nostrils, big overlapping eyelid scales, and skin that covers the ears. It is really fun to watch them zoom across a sand dune and quite literally dive into the sand. You should visit one of the dunes where these lizards live and hike around to find them. Since they only occur in habitats associated with loose sands, and since there are so few of these habitats, the Coachella Fringe-toed Lizard is endangered, and the other two species are protected as species of special concern. Development has impacted the Coachella Fringe-toed Lizard in particular because the majority of its historic range has been converted into farms or housing developments. Off-road vehicle traffic on sand dunes has exacerbated the lizards' problems. Watch the lizards from afar by eye or via binoculars. Catching and handling fringe-toed lizards is prohibited in California without special permits.

Coachella Fringe-toed Lizard (juvenile). *Photograph by Alex Mason.*

Appearance: Fringe-toed lizards have flattened bodies with legs that extend out to the side so that they maintain a low center of gravity when zooming over the sand. They have a mottled white and brown pattern on their back that up close resembles dozens of tiny white flowers with reddish-brown markings set against a dark background. They have sharp, upturned duckbill-shaped snouts. The best way to identify which species of fringe-toed lizard you are observing is based on your location (see Range and Variations section below). However, here are some differences among the three. The Coachella Fringe-toed Lizard (*Uma inornata*) typically lacks the black mark on each side of the belly possessed by the other two species, and if they do have marks, they are small and faded. The Colorado Desert Fringe-toed Lizard (*Uma notata*) has a distinct black and bright-orange mark on each side of the belly. The Mohave Fringe-toed Lizard (*Uma scoparia*) has a black mark on each side of the belly, with dark, boomerang-shaped marks on the chin, and the dark marks on its back do not align to form stripes like they do in the other two species.

Natural History: Fringe-toed lizards are adapted to living in loose sands, and so can be found at the edges of sand dunes in areas with sparse vegetation, as well as sandy bajadas and sometimes in large washes. The fringes on their toes facilitate rapid running across sand, and their duckbill-shaped snouts allow them to bury themselves in the sand. They are active during the day, when they run around ambushing insects on the sand and plucking them from desert vegetation. Adult lizards can occasionally eat small lizards. The timing of the reproductive cycle varies a bit by species, but in general, they mate in the spring, lay eggs in the summer, and hatchlings appear in late summer and early fall.

Range and Variations: The endangered Coachella Fringe-toed Lizard lives in isolated pockets of sandy areas in the Coachella Valley in Riverside County. The Colorado Desert Fringe-toed Lizard lives in sand dunes and other sandy areas in Imperial County and extreme northeastern San Diego County and extends a short distance into northern Baja California. The Mohave Fringe-toed Lizard is the most widespread of the three species, although they are still restricted to sand dunes and certain other sandy areas. They range through much of San Bernardino County and eastern Riverside County. Small populations have also been documented in extreme western Arizona and southern Nevada.

How to Find Fringe-toed Lizards: Given that you are not allowed to capture these lizards, go to sandy areas where they live and admire them from a distance. Avoid driving over sand dunes because this contributes to their destruction, which is one reason fringe-toed lizards are at risk. The easiest way to find them is to walk around the edges of sand dunes, where vegetation first appears, on warm days in the spring. They blend into the sand very well, so you might only notice one when it zooms across the sand at your approach. Bring binoculars to scan the sand dunes and to observe the lizard's features.

Long-tailed Brush Lizards use camouflage to avoid detection by predators.

BRUSH LIZARDS

UROSAURUS GRACIOSUS AND *U. MICROSCUTATUS*

FAMILY PHRYNOSOMATIDAE

These plain little desert denizens are not at the top of most lizard hunters' wish lists. They might be small and brownish, but their seemingly nondescript appearance was put into a different light for me by a brilliant mentor. I was lucky to be an undergraduate curatorial assistant in UC Berkeley's Museum of Vertebrate Zoology in the late 1990s. My job was to preserve and catalogue the specimens that scientists brought in to be deposited forever in the museum, and to help visiting scientists retrieve specimens from the collection for study. What a dream job for an aspiring herpetologist! The 1990s were an especially exciting time because, in addition to all the excellent herpetologists employed by the museum at that time, I got to hang around with Dr. Robert Stebbins, the very first curator of herpetology at the museum. He was long since retired by this time, but he never stopped bringing in specimens and painting their likenesses for his renowned *Field Guide to Western Reptiles and Amphibians*. He shared many stories with me over the two years I was there, and he would eagerly listen to me tell him about the creatures I encountered on my trips to the desert. Dr. Stebbins was the one who introduced me to Long-tailed Brush Lizards, a species that is incredibly common in the Mojave Desert but that most visitors never see. He told me about how these lizards live on the branches of creosote bushes and the pattern on their skin causes them to blend right in, such that you won't see them unless you are looking for them. On my next trip, I eagerly scanned the branches of bushes, and had no luck and was about to give up after an hour or so when . . . voilà! The outline of a little lizard appeared on one of the bushes! Their camouflage is truly amazing. I know now how privileged I was to get these tips from one of the most famous herpetologists of all time, and I am glad to pass them on to you here.

Appearance: The Long-tailed Brush Lizard and Small-scaled Brush Lizard are very similar in appearance. The main way to tell them apart is based on location, but they do overlap in a few areas in eastern San Diego County. Long-tailed Brush Lizards (*Urosaurus* *graciosus*) are small, thin lizards with pointy snouts and with tails that are about twice the length of the body. In Small-scaled Brush Lizards (*Urosaurus microscutatus*), the tail is not quite as long and is often black. Both species are grayish with faint dark markings that make their skin resemble the pattern of the branches on which they perch. Their scales are small and granular, with a stripe of larger scales that extends down the center of their backs. These central scales are larger in Long-tailed Brush Lizards than in Small-scaled Brush Lizards. Males have beautiful blue or green patches on the belly, and both sexes can sometimes have yellow or orange markings on the chin.

Natural History: Long-tailed Brush Lizards are a desert species; they live on creosote bush and other arid-adapted shrubs. Small-scaled Brush Lizards are typically found on small rocks in desert habitats. Both species eat small insects and other invertebrates. They mate in the spring, females lay one or two clutches of eggs in the summer, and hatchlings appear in the summer and early fall.

Range and Variations: Long-tailed Brush Lizards live in the southeastern deserts from San Bernardino County southward into Baja California and northern mainland Mexico. They also occur in southern Nevada and western Arizona. Small-scaled Brush Lizards are mainly a Baja California species, and their range extends into the United States in eastern San Diego County and tiny parts of western Imperial County and southern Riverside County.

How to Find Brush Lizards: Follow Dr. Stebbins's advice and look for Long-tailed Brush Lizards lying flat against the branches of creosote bushes and other shrubs or small trees. The best bushes to look in are big ones with complex root systems where the lizards can hide at night. Notably, these lizards sometimes sleep on the branches too; I have found them by scanning creosote with my flashlight at night. You can also find these lizards climbing on wood fences in appropriate habitat, which can be much easier than searching for the camouflaged little buggers on creosote. For Small-scaled Brush Lizards, you should hike around in rocky areas with plenty of vegetation and look for small lizards basking on rocks or branches.

Jeff Lemm

This Small-scaled Brush Lizard is doing a push-up, a behavior meant to communicate with other lizards about territorial boundaries.

An Ornate Tree Lizard preying upon a bee. *Photograph by Jeff Martineau.*

ORNATE TREE LIZARD

UROSAURUS ORNATUS

FAMILY PHRYNOSOMATIDAE

The Ornate Tree Lizard occupies much of the southwestern United States, but its native range just barely extends into California. They are extremely common in Arizona and New Mexico, where their biology has been well-studied. They exhibit fascinating variations in the color of their throat patches, especially in males, both among and within populations. Throat color appears to relate to social status, as males from Arizona with large blue spots on their throats tend to win fights against males with smaller spots. A study in New Mexico demonstrated that males with green spots on their throats were dominant to orange-throated males. So, when it comes to social dominance hierarchies in Ornate Tree Lizards, it's complicated. Why chin color? These colors often reflect differences among males in physiology and mating strategy, where one color might be beneficial under certain conditions and a different color works best under others. The males can flash their chins at one another, and at females, and they use their excellent vision to make assessments. It's not too different from how people look out across a dance floor to take stock of potential mates and competitors!

Appearance: Ornate Tree Lizards are small lizards with long tails, though their tails are not as long as those of the aptly named Long-tailed Brush Lizard. Their color varies dramatically from light gray to black, and individuals can change their color when stressed, due to temperature, or for other reasons. Their scales are small and granular, and down the centers of their backs they have two rows of slightly enlarged scales with small scales between them (another way to distinguish them from other species in the genus *Urosaurus*). Males have blue or green belly patches, and both sexes have chin colors that vary among populations.

Natural History: Ornate Tree Lizards certainly live on trees and shrubs, but they can also be found on rocks or on man-made structures like fence posts. They are active during the day when they bask in the sun, and at night they retreat under rocks, into burrows, or into tree roots or bark. They eat insects and arachnids. They mate in the spring, and females lay multiple clutches of eggs throughout the summer, so hatchlings can emerge anytime in the summer and fall.

Range and Variations: In California, Ornate Tree Lizards are found along the Colorado River in the extreme eastern parts of San Bernardino, Riverside, and Imperial Counties. There are also several populations of Ornate Tree Lizards that have been introduced farther west, including Barstow, San Bernardino, and El Centro, where they occupy small areas. Scientists regularly discover Ornate Tree Lizards in new areas of California, suggesting that they might be dispersing by hitching rides with people. From eastern California, the natural range of Ornate Tree Lizards spreads eastward into central Texas, north to southern Colorado, and south to northern Mexico.

How to Find Ornate Tree Lizards: In California, you can find Ornate Tree Lizards perched on trees or rocks alongside the western edge of the Colorado River. Walk around near the riverbank at public access points looking for small lizards basking in the sun on warm days. You might also encounter these lizards in one of their many introduced populations. Ornate Tree Lizards are easy to capture via lasso. Doing so allows you to admire their beautiful throat colors up close and personal before letting them go.

COMMON SIDE-BLOTCHED LIZARD

UTA STANSBURIANA

FAMILY PHRYNOSOMATIDAE

The small, unassuming Common Side-blotched Lizard has the distinction of being California's most studied lizard. Every other lizard species probably has just as much fascinating stuff to be learned about it, but scientists have unearthed more about the abundant and charismatic little Common Side-blotched Lizard than any other to date. A good chunk of this comes from work done by Dr. Barry Sinervo and his colleagues and students, who studied polymorphism in Common Side-blotched Lizards in detail. *Polymorphism* means "many forms," and in this case it refers to the fact that the males in a population these scientists studied for many years have one of three throat-patch colors: blue, yellow, or orange. These colors are genetically determined, and the three male "morphs" behave very differently when it comes to the lady lizards. Orange males have big territories with multiple females in them, and they patrol these territories aggressively, fighting off other males. Blue males have small territories with one female only. Yellow males don't have a territory at all, but instead roam around trying to mate with females on other males' territories. In this way, all three male morphs coexist because each one wins against one other morph and loses against one other morph in a real-life rock-paper-scissors game: orange beats blue because they have more females in their territory, yellow beats orange because they can sneak a tryst with a female when the male is elsewhere on his huge territory, and blue beats yellow because the males guard their females and prevent the yellow males from sneaking in. There is so much more to the story—a whole book could be written on the fascinating mating systems of Common Side-blotched Lizards. One thing is for sure: these lizards' diminutive size, plus their abundance and how widespread they are in California, often makes people overlook them in search of rare or hard-to-find lizards. But if you just stop to take a close look at the next Common Side-blotched Lizard you see, especially a showy male, you might be amazed at its beautiful, intricate pattern, its bright coloration, and the hilarious yet noble shuddering display, along with the classic side-eye lizard glare it delivers when you get too close to it.

Appearance: Common Side-blotched Lizards are small gray or brown lizards with highly variable patterns and colors that sometimes form stripes, bands, little chevrons, or no pattern at all. Their bellies are pale, with the notable exception of the namesake dark blotch on each side of the abdomen, which confusingly is absent in some individuals. Males often have showy colors, including a blue, orange, or yellow chin, plus blue spots all over the back, or many possible other colors and patterns that are too numerous to describe here. Though the structure of their scales varies across their bodies, overall the scales are small and smooth, making it difficult to mistake adult Common Side-blotched Lizards for juvenile, spiky-scaled lizards of the genus *Sceloporus*.

Natural History: Common Side-blotched Lizards occupy many habitats in California, but most of these areas have one thing in common: small rocks. They are found in desert, grassland, chaparral, even rural and suburban yards, as long as there is plenty of open space to allow the sun to shine through. Common Side-blotched Lizards bask in the sun on small rocks or on the ground during the day and hide under rocks or in burrows at night. They eat small insects and arachnids. They mate in the spring, lay eggs in the spring and summer, and hatchlings appear in summer and fall.

Range and Variations: Common Side-blotched Lizards live in most of California, from the Bay Area southward, plus they dip into the desert areas of the extreme northeastern part of the state. They are absent from most of the Sierra Nevada. Outside of California, Common Side-blotched Lizards extend from central Washington southeast into central Texas and northern Mexico, as well as into all of Baja California.

Jeff Martineau

How to Find Common Side-blotched Lizards: Common Side-blotched Lizards are one of the most common and conspicuous lizards in California. It is difficult not to see them if you are in their habitat! Go for a hike on a warm day in any habitat where they occur, and you will see many Common Side-blotched Lizards basking on rocks, on the dirt, or occasionally on a piece of wood or a low branch. But just because they are easy to find doesn't mean they are easy to catch! Common Side-blotched Lizards allow you to approach them but can be lasso-dodgers to the fullest degree.

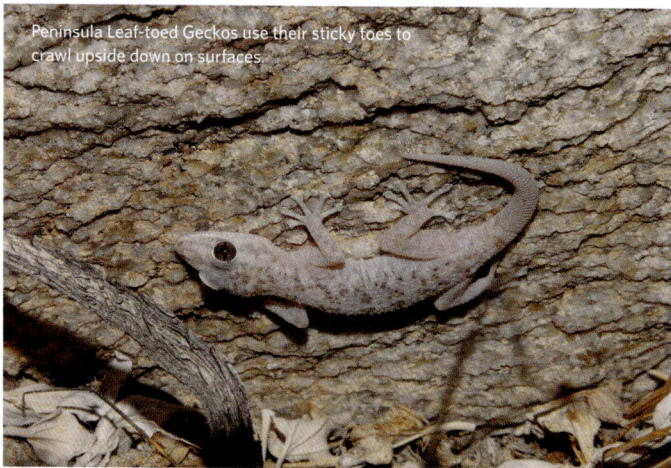

Peninsula Leaf-toed Geckos use their sticky toes to crawl upside down on surfaces.

Jeff Lemm

PENINSULA LEAF-TOED GECKO

PHYLLODACTYLUS NOCTICOLUS

FAMILY PHYLLODACTYLIDAE

If by chance you have read the species accounts in this book in order, you may recognize a common pattern by now: many lizards occur mainly in Baja California and just barely peek into Southern California, typically into San Diego, Imperial, and Riverside Counties. As discussed in the species account for the Baja California Collared Lizard (page 62), this is because the Baja California peninsula broke off from mainland Mexico millions of years ago and slowly moved northward as an island until it merged with Southern California, bringing all its animals with it. The

A group of **Peninsula Leaf-toed Geckos**. *Photograph by Jeff Lemm.*

Peninsula Leaf-toed Gecko is a great example of such a species. These lizards are the only species from the family Phyllodactylidae that occur in the United States, while dozens of other species in the genus *Phyllodactylus* live throughout Mexico, Central, and South America. They are closely related to geckos in the family Gekkonidae, and the two groups share numerous characteristics like lack of moveable eyelids and the propensity to climb around on hard surfaces at night. However, because the Peninsula Leaf-toed Gecko is in their native range in California, we get to see them living on natural substrates like desert boulders instead of on buildings.

Appearance: Peninsula Leaf-toed Geckos are small lizards that are typically pale pinkish gray, with a light mottling of brown spots. The skin is composed mainly of tiny, smooth scales with occasional enlarged bumps. Like their relatives in the family Gekkonidae, they have large eyes with no eyelids, and their pupils are vertical. The most obvious way to distinguish them from other geckos is the shape of their toes, which at the tips splay out into little, lobed toe pads that somewhat resemble leaves.

Natural History: Peninsula Leaf-toed Geckos live in large boulder fields in the desert and chaparral, where they hide in rock crevices and under flat rocks during the day and are active at night. After the sun goes down, they climb around on large boulders, where they capture small insects and arachnids. They mate in the spring, females lay multiple clutches of eggs during the summer, and eggs hatch later in the summer.

Range and Variations: The Peninsula Leaf-toed Gecko occupies much of central Baja California, with a narrow strip of its geographic range extending northward into the desert areas of eastern San Diego, western Imperial, and western Riverside Counties.

How to Find Peninsula Leaf-toed Geckos: The most difficult thing about finding Peninsula Leaf-toed Geckos is that they occur in relatively remote areas, so you need to plan your trip carefully. They are plentiful in rocky areas, especially those with lots of interspersed vegetation and creek beds. Go for a night hike with a good flashlight and watch for lizards running around on large boulders.

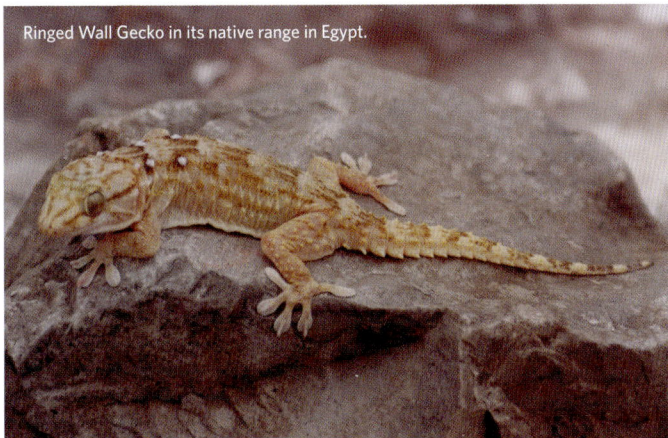
Ringed Wall Gecko in its native range in Egypt.

Adel Ibrahim

WALL GECKOS

TARENTOLA ANNULARIS AND *T. MAURITANICA*

FAMILY PHYLLODACTYLIDAE

It's unfortunate that these chubby, gorgeous lizards are not native to California, just hanging around on houses all over the place for us to admire. Alas, they are introduced to the state, as are all lizards in the family Gekkonidae, and are only present in isolated neighborhoods near where they were first released. Of course, non-native species are *not* a good thing because they can out-compete native lizards and potentially spread disease. But boy howdy, these chonky lizards are cute and charismatic! Both species have become established in California as the result of escaped pets. Like the geckos in the genus *Hemidactylus* (page 84), they climb around on the outside of buildings where they can be seen squabbling for the best spots under light fixtures, sometimes making audible squeaks at one another.

Moorish Gecko.

Appearance: These medium-sized geckos are stout-bodied, with enlarged toe pads, big eyes with vertical pupils, and prominent bumps on their skin. The somewhat larger Ringed Wall Gecko (*Tarentola annularis*) is a mottled light brown, and adults have four little white marks surrounded by dark pigment on their upper backs. The slightly smaller Moorish Gecko (*Tarentola mauritanica*) is similarly light brown but has mottled dark and light spots all over its body, and the tails of adults appear banded. Moorish Geckos can be active day or night, and their color can darken during the daytime to help them heat up from solar radiation.

Natural History: Both species of these geckos can be found in isolated populations on houses and alleyways in urban areas. They are active primarily at night, feeding on insects and the occasional hatchling lizard. Unlike most geckos, Moorish Geckos can be found basking in the sun, usually in the morning when it is not terribly hot outside. Little is known about their reproduction, except that females lay several clutches of one or two eggs over an extended period in the summer.

Range and Variations: The Ringed Wall Gecko is native to northern Africa, and introduced populations exist in Florida, Arizona, and in California, where they are established in two known populations: the cities of Redlands and San Juan Capistrano. The Moorish Gecko is native to northern Africa and southern Europe, and has been introduced to Florida, Texas, several areas in South America, and many places in Southern California, most notably El Cajon in San Diego County.

How to Find Wall Geckos: These lizards are not very easy to find, especially the Ringed Wall Gecko, which only occupies two small urban areas to my knowledge. Information on how to find the exact neighborhoods that the lizards inhabit is not readily available, so you would need to play detective to try to figure out where they are. Searching for these lizards involves the same strategy as looking for *Hemidactylus* lizards: use a flashlight at night to search for the lizards sitting on walls. Be careful not to confuse *Hemidactylus* geckos with these much larger, chubbier *Tarentola* geckos. Juveniles can be especially difficult to identify. If you get a decent photograph, then you can post it to iNaturalist, and experts can confirm the species. You will also be helping biologists learn about the spread of these introduced species!

Jeff Lemm

OCELLATED SKINK

CHALCIDES OCELLATUS

~~~~~~

FAMILY SCINCIDAE

I already had a complete draft of this book written when I found out about California's newest introduced species, the Ocellated Skink. There's nothing like observing a species invade the state in real time to bring home how timely and dire the issue of exotic species introduction is. It's not just lizards, it's the non-native plants taking over our grasslands and hillsides; it's the introduced freshwater mussels choking our lakes and reservoirs; it's the invasive moths that killed off most of the official tree species of the city of Los Angeles (ironically a non-native plant species itself); it's the Argentine ants slowly retracting the range of the Coast

Horned Lizard (see page 106). It's a mess, is what it is. So, how did this latest invader make its way from its native range in southern Europe and northern Africa to our similarly Mediterranean climes? Likely the same way so many other lizards have become introduced in Southern California: via the plant nursery and pet trades. Genetic tests are underway, but scientists hypothesize that the populations of Ocellated Skinks popping up in Southern California probably jumped here recently as stowaways from a large tree nursery in Arizona, where this species is established. They were also popular in the pet trade several decades back, making it possible that the initial American stock arose from released pets, then spread via plant root balls.

Appearance: These smooth, shiny lizards are small or medium in size, have pointy noses and no discernable neck, with small limbs and a long, thick tail. They are pale underneath and vary in their coloration on top but are typically golden with little dark spots that align to form jagged bands. The dark spots have tiny white vertical stripes through them, reminiscent of the Eye of Sauron. In fact, the root word *ocelli* means "eyespots."

Natural History: Nothing is known about the Ocellated Skink's habits in California, given that their introduction was so recently discovered. In their native ranges and in introduced populations in Arizona, these skinks are active during the day in a wide variety of habitats, where they skulk between sunny areas and cover. In California, they are found mainly in people's yards. These skinks eat insects and arachnids, and large individuals also eat small lizards. They mate in the spring and females give birth to one or two clutches of live babies in the summer.

Range and Variations: The native range of the Ocellated Skink is southern Europe and northern Africa. It is also found in parts of southern Asia, where it was likely introduced, as well as numerous other locales throughout the world, including Florida and Arizona. It has recently been discovered in San Diego and San Bernardino Counties and could be present in other areas in Southern California as well.

How to Find Ocellated Skinks: As is true for some other non-native lizard species in California, the Ocellated Skink has a limited geographic distribution in Southern California (for now) due to its presumably very recent introduction. Scientists are systematically searching for the skinks in areas where they've been reported in an effort to determine whether they could be eradicated before they become permanently established. These areas have not all been made public. If you think you might be in an area near these populations, you can search for the skinks by walking around on a warm day and watching for shiny lizards darting for cover under plants, or by flipping cover objects to search for them. If you find what you think is an Ocellated Skink, please post your observation to iNaturalist.

GILBERT'S SKINK

PLESTIODON GILBERTI

FAMILY SCINCIDAE

Where I live, near the coast in the San Luis Obispo area, we have the Western Skink, but not the Gilbert's Skink. Finding the larger, beefier Gilbert's Skink requires driving an hour or so inland, over the Coast Ranges where Western Skinks no longer occur and Gilbert's Skinks emerge. I teach herpetology each spring, and we do plenty of local field trips during class hours where students observe Western Skinks in the wild. I also offer an optional field trip to the Carrizo Plain on a Saturday each spring, where my students and I explore a private ranch that is brimming with

Juvenile Gilbert's Skink. *Photograph by Chad Lane.*

herps, including the Gilbert's Skink. Dubbed "beefcakes" by those students who are dedicated enough to go on the optional weekend class field trip, Gilbert's Skinks are a real treat to meet in person. In spring, the big males are out on the prowl, slinking through the brush and showing off their bright salmon mating coloration. If we are lucky enough to catch one of these wary beasts to admire it up close, care must be taken not to get bitten, especially by the big, strong-jawed males. One intrepid herpetology student had a habit of letting any lizard he captured bite onto his earlobe, where it would then hang like an earring for a moment as he took a selfie.

He did this with Great Basin Collared Lizards, Southern Alligator Lizards, and Gilbert's Skinks, and said that the skink packed the hardest bite. I do NOT recommend such antics, as sooner or later someone could lose an earlobe. Indeed, the bite mark left behind by this stunt did demonstrate just how strong the bite of a big, beautiful Gilbert's Skink can be.

Appearance: Adult Gilbert's Skinks are large lizards with smooth, shiny scales, narrow heads with no discernable neck, and long tails. They are typically tan, gray, or olive green, but large lizards can take on an orange coloration, especially males during the spring mating season. Juveniles have dark and light stripes down their backs that fade with age, plus bright-orange or blue tails depending on their location. In some areas, hatchling Gilbert's Skinks can sport very attractive bright-red tails and legs.

Natural History: Gilbert's Skinks inhabit grasslands, woodlands, and moist areas in chaparral. These lizards spend much of their time underground or under cover objects. They only bask in the sun for short periods of time to warm up, and most of the active time above ground is spent lurking around under shrubs or near the mouths of their burrows. Gilbert's Skinks eat many types of invertebrates that inhabit the soil and leaf litter. They mate in the spring and summer, females lay eggs in the summer and stay with the eggs to protect them, and the eggs hatch in the fall.

Range and Variations: Gilbert's Skinks occur throughout much of central California and Southern California, from Yuba County southward, but they do not quite reach the coasts. In deserts, they can only be found at high elevations in desert mountain ranges. They occur in the foothills of the Sierra Nevada at low and mid elevations, as well as a patchy distribution in Southern California mountains and in several mountains and valleys east of the Sierra Nevada. They also occur in southern Nevada, western Arizona, and northern Baja California.

How to Find Gilbert's Skinks: During the spring, you might observe Gilbert's Skinks active on the surface. I often see a silver-orange flash out of the corner of my eye as a skink dives for cover under a rock or into its burrow. Your best bet for finding Gilbert's Skinks is to pretend as though you are looking for snakes, and flip cover objects, including natural ones like flat rocks or logs, or man-made ones like pieces of tin, plywood, or carpet at junk piles. For Gilbert's Skinks, I have much more luck at junk piles than I do with natural cover objects. Be sure to always replace all cover objects, natural or man-made, exactly as you found them. Skinks are best captured by hand, but be careful not to touch their tails so they don't drop them. Lassoing skinks is not really practical because they are seldom sitting out in an accessible place, but also because they don't really have necks and so a lasso could slip right off.

Max Roberts

WESTERN SKINK

PLESTIODON SKILTONIANUS

You might have heard of blue-tongued skinks, rather large beasts from Australia that are popular in the pet trade. Well, in California, we have our own versions of such lizards, except it's the tails that are blue. Whether on tongues or tails, the blue helps protect against predators. When a Blue-tongued Skink is confronted by a predator, it opens its mouth wide and sticks its blue tongue out at the offending snake or bird, which looks comical to us but apparently actually works to ward off some attacks. The blue tongues also reflect ultraviolet light, which is invisible to us but helps scare off these predators. In the case of the Western Skink, also

Marisa Ishimatsu

Juvenile Western Skinks have bright-blue tails.

known as the Skilton's Skink, the juveniles have uber-conspicuous bright-blue tails that reflect ultraviolet light as well. These colors make the tails especially obvious to predators, so that they attack the tails instead of the body. It seems like quite a risk for a skink to have a bright-blue tail that *attracts* predators, but it must be working because blue tails are very common among young skinks and indeed many other lizards like whiptails. The blue can even remain faintly in some adults. When the predator goes for the tail, it breaks right off, and the tail wiggles like crazy for a few minutes to distract the predator while the rest of the skink slinks away to safety. They then grow the tail back over the next month or two, but interestingly, they cannot regrow bone. The new tail has a cartilaginous rod inside it instead of vertebrae, and so the regrown part cannot break again.

Appearance: Western Skinks are small to medium-sized lizards with distinctive black, brown, and cream stripes down their backs, smooth and shiny scales, and a long tail that is bright blue in juveniles and pale blue or gray in adults. They have pointy snouts and lack a noticeable neck.

Natural History: Western Skinks are very common lizards in multiple habitats in California, including forests, grassland, and chaparral. They are especially common in areas with rocks among the vegetation and in areas near ponds or creeks. They are active during the day but spend a lot of time underground or skulking under cover objects or in the leaf litter. You are much more likely to see a flash of their blue tail disappearing under a rock or into a burrow than you are to see a Western Skink out basking. They consume arthropods that live in the leaf litter and soil. Western Skinks mate in the spring, earlier in southern populations than in northern populations, and lay their eggs in the late spring or summer. Females stay with the eggs and protect them from predators. Hatchlings can be found in summer and fall.

Range and Variations: The Western Skink has a wide range in California, spanning the coastal counties, much of Northern California and the northern Sierra Nevada, plus several populations in the southern Sierra Nevada. It also ranges through Oregon, Washington, Idaho, and western Montana into southern British Columbia, plus Nevada, western Utah, northern Arizona, and northern Baja California.

How to Find Western Skinks: Western Skinks are difficult to find in many places because they spend so much time underground or under cover objects. While you certainly can hike around in appropriate habitat to look for them, keeping your eyes out for small flashes of shiny striped scales and blue tails, your best bet is to search under the cover objects that they frequent. I have good luck by flipping flat rocks in grasslands and open areas of woodlands, especially those near bodies of water. Western Skinks can also be found under logs sometimes, and if you are lucky enough to find pieces of plywood or tin or other junk items that attract reptiles, give those a flip, too. Always be sure to put the cover objects you flip back exactly as you found them.

Jeff Nordland

ORANGE-THROATED WHIPTAIL

ASPIDOSCELIS HYPERYTHRUS

FAMILY TEIIDAE

Whiptail lizards never stop moving. They are unique among California lizards in this way. Whereas other daytime-active lizard species can be categorized as sun baskers or shade lurkers, the whiptails move too much to fit either definition. They are active when it is very hot out and spend plenty of time in the sun. However, they continually move from shrub to shrub with their signature jerking motions, never sitting in one place to bask for too long. What are they doing when they jerk across the ground? Foraging. These lizards are constantly flicking their snake-like forked tongues in search of bugs in every nook and cranny. Whiptails can also run really fast, which they are glad to do if they get spooked. Orange-throated Whiptails are one of the most beautiful species of whiptail lizard, especially the brightly colored males during the mating season. Native to extreme Southern California and the Baja California peninsula, these lizards used to be common in their range north of the US–Mexico border, but the massive number of housing developments built in San Diego, Orange, and Riverside Counties is threatening this species. While still common in some areas, they are on the California Department of Fish and Wildlife's watch list, meaning that more information is necessary to discern whether they should be named a species of special concern.

Appearance: Orange-throated Whiptails are medium-sized, slender lizards with pointy snouts, large flat scales adorning the top of the head, and tails that are twice as long as the body. These lizards have multiple dark and light stripes extending down their backs and onto the tail. The top surface of their body has tiny granular scales, and their bellies consist of rows of rectangular scales. Females have white undersides which sometimes become slightly orange during mating season, while in males the chins and sometimes even the rest of the undersides are an extremely bright orange.

Natural History: Orange-throated Whiptails inhabit dry areas with vegetation including chaparral and desert washes. They are active during the day, often when it is very hot out. They spend most of their time moving through the environment searching for termites and other arthropods at the bases of shrubs and rocks, poking their long noses into leaf litter and loose soil, and digging with their front legs. Orange-throated Whiptails mate in the spring, females lay eggs in the summer, and hatchlings appear in late summer.

Range and Variations: Orange-throated Whiptails range in coastal and near-coastal areas from Orange County and southwestern San Bernardino County south through western Riverside and San Diego Counties, all the way down to the tip of Baja California.

How to Find Orange-throated Whiptails: Go hiking in chaparral or on hillsides with rocks and sparse vegetation on a warm day in the spring or summer, and you will likely see Orange-throated Whiptails out foraging. You might see a similar species, the Western Whiptail (page 173), but these lizards have mottled spots on their backs instead of stripes. Trying to capture whiptail lizards is a challenge. They are very wary, are fast enough that it is difficult to hand catch them, and they rarely sit still long enough for you to succeed with a lasso.

Marisa Ishimatsu

Juvenile Orange-throated Whiptail.

Sonoran Spotted Whiptail.

NON-NATIVE WHIPTAILS

ASPIDOSCELIS SONORAE AND *A. TESSELATUS*

FAMILY TEIIDAE

The Sonoran Spotted Whiptail and Common Checkered Whiptails are not native to California but have been introduced to Southern California under extraordinary circumstances. Just like the Indo-Pacific House Gecko, which has taken up residence in the same part of the state, these two whiptail lizards are all-female species that reproduce without males. They are native to other areas in the southwestern United States, but apparently people released a Sonoran Spotted Whiptail in Orange County and a Common Checkered Whiptail in Los Angeles County sometime within the last decade or two, whether on purpose or by accident. Just think about it—if there are no males, then a single female can easily establish a local population, and if conditions are good for that species, then they will just keep on spreading. This issue was exacerbated by the fact that these species resemble native whiptails, so many people didn't notice that these animals were not a native species. Even images on iNaturalist were misidentified until scientists took a closer look at some of the records. Records of these species or possibly of other non-native whiptail species have popped up in different areas of the state, but the extent to which these are from established populations remains unclear.

Appearance: Sonoran Spotted Whiptails (*Aspidoscelis sonorae*) resemble native Orange-throated Whiptails in that they are medium-sized, thin lizards with pointy noses, stripes down their backs, and very long tails. One difference is that they have tiny faint or bold spots speckling the dark stripes on their backs. Their undersides are pale. Common Checkered Whiptails (*Aspidoscelis tesselatus*) are similar in size and highly variable in color, but generally they have a dark checkered pattern that can sometimes merge into stripes, on a lighter background. Their thighs are often dark with bold light spots on them.

Natural History: In California where they are introduced, these whiptails thrive in yards, parks, and parking lots in urban areas. They are active during the day, where they forage for insects and arachnids amid landscaped yards and take shelter in crevices or underneath concrete. In these all-female species, individuals lay several clutches of eggs during the summer that hatch a couple of months later.

Range and Variations: Sonoran Spotted Whiptails are established in parts of Orange and San Diego Counties and appear

Patrick Alexander

Common Checkered Whiptail.

to be spreading rapidly. Several individuals of what might be a population of Sonoran Spotted Whiptails have also been seen in the vicinity of Sacramento. Their native range includes Arizona, western New Mexico, and northern Mexico. Common Checkered Whiptails appear to have had a single recent introduction into Los Angeles County from their native range in Colorado, New Mexico, Texas, and northern Mexico.

How to Find Non-native Whiptails: You can search for Sonoran Spotted Whiptails by walking around in neighborhoods, parking lots, and parks in areas where they have been previously spotted (you can check iNaturalist to see an up-to-date map of verified locations). Look for long, thin lizards that walk around on the ground with a jerky motion, poking their heads into the soil and scrabbling about in the leaf litter. Although it can be difficult to train binoculars or your camera on these lizards because they so rarely stop moving, that is the easiest way to look for the markings that can identify them. Be sure to post photos of these lizards to iNaturalist for confirmation of the ID and so that scientists can continue to study their spread.

Chad Lane

WESTERN WHIPTAIL

ASPIDOSCELIS TIGRIS

Known by many as "speedos" due to their zoominess and our love for scientific-name wordplay (the genus can be pronounced "aa-*speedo*-selis"), Western Whiptails are commonly encountered running rapidly across our paths in many California habitats. These lizards, also known as Tiger Whiptails, are my nemesis. While I love these fascinating lizards dearly, they are extremely difficult to see up close and even more difficult to capture. I recently did a research project with my graduate student that required capturing Western Whiptails and numerous other

Max Roberts

Mojave Desert lizard species to measure their hydration and evaporative water loss across the skin. We spent *hours* trying to capture these lizards! It's not just us, trust me. These lizards are famously difficult to catch. When I was in college, the graduate students whispered of a legendary event that had taken place on previous herpetology class field trips. To understand this, you need to know about two words that were in use back in the day. Whiptails used to be in the genus *Cnemidophorus*, which has a silent *c*, and folks referred to the lizards as "cnemis" (pronounced "nem-eez"). Also, people used to refer to lassoing lizards as "noosing" them. So, here is the story. Allegedly, classes in the past had formed the Naked Cnemi Noosers club, where students would find a Western Whiptail, strip naked, and work together to catch a lizard. They could not put their clothes back on until someone had the lizard in hand. Given my experience with Western Whiptails, I am willing to bet that those students racked up quite a few pokes from sharp desert plants and severe cases of sunburn in pursuit of noosing a cnemi while naked.

Appearance: Western Whiptails share a body plan with other whiptail lizards, including medium size, long tails, and pointy snouts. They have smooth, granular scales on their back, with several different colors ranging from tan to yellow to brown forming mottled bars, spots, or irregular stripes. Juvenile lizards have a striped pattern and blue tails. Western Whiptails have rectangular belly scales arranged in rows, and the scales on their tails are rather spiky.

Natural History: Given the wide range of Western Whiptails in California, it makes sense that they occupy many habitat types, including arid deserts and chaparral, along with open woodlands. They avoid areas with dense plants, so are not typically found in grasslands or thick forests. Western Whiptails are highly active, moving about all day in their signature jerking fashion as they forage for insects and arachnids underneath plants, and occasionally sprinting to capture an arthropod or small lizard in the open. They tend to be active when it is quite hot out. They mate in the spring and summer, lay eggs shortly thereafter, and hatchlings appear in the late summer and early fall.

Range and Variations: Western Whiptails are present throughout most of California from Trinity and Shasta Counties southward but are absent from the north coast and from much of the Sierra Nevada. Outside of California, they occupy the Great Basin and Sonoran Deserts, ranging from Oregon and Idaho to western Colorado and New Mexico and south into mainland Mexico and Baja California.

How to Find Western Whiptails: In many areas, such as the California deserts, Western Whiptails are incredibly common. They are easiest to find in flat areas with intermittent vegetation, so hike around in washes or creosote flats on a hot day and you are likely to see one of these lizards within minutes. If you drive on desert roads during the day, you may find yourself dodging these lizards as they dash across. In other areas, they can be more sparsely distributed and a bit harder to find. I live near San Luis Obispo, at the edge of the Western Whiptail's range where they approach the coast. Here we only see whiptails when it is hot out, and only infrequently at that.

Granite Night Lizard.
Photograph by Zeev Nitzan Ginsburg.

SANDSTONE AND GRANITE NIGHT LIZARDS

XANTUSIA GRACILIS AND *X. HENSHAWI*

FAMILY XANTUSIIDAE

These two closely related native species live up to their names by being active at night, especially during the summer. They are similar to the non-native geckos that inhabit the California deserts, in terms of their small size and their unlidded eyes with vertical pupils, but the Granite and Sandstone Night Lizards climb around

Sandstone Night Lizard.

on rocky outcrops instead of on buildings. Granite Night Lizards are rather common, but most people in California will never see them because they occupy only the remotest reaches of the desert in the extreme southern part of the state. As with several other species like the Baja California Collared Lizard, the Granite Spiny Lizard, and the Banded Rock Lizard, these night lizards do not extend north of the San Gorgonio Pass (where Interstate 10 crosses through the valley north of the San Jacinto Mountains), likely because they only arrived in Southern California when the then-island of Baja California merged with the mainland 10 million years ago. The two species were considered one and the same until about twenty years ago, when a study revealed that the Sandstone Night Lizard was different enough to be named its own species despite being restricted to a tiny geographic range. As their names suggest, Granite Night Lizards live underneath flakes of granite rocks, while Sandstone Night Lizards take refuge in crevices and holes in sandstone. While they are similar in appearance, studies have revealed differences in head, body, and limb size and shape that resulted from natural selection in these respective habitats.

Appearance: These small lizards have a yellowish-gray background color with large dark spots all over. The Sandstone Night Lizard (*Xantusia gracilis*) always maintains this coloration, while the Granite Night Lizard (*Xantusia henshawi*) turns darker during the day. Their flattened heads are covered in wide scales, but the scales on their backs are tiny and granular, giving them a fine, bumpy appearance.

Natural History: These lizards are restricted to areas with their namesake rocks. Sandstone Night Lizards are typically only active at night. Granite Night Lizards are also out and about at night during warm times of year, but they exhibit some daytime activity, both underneath the granite flakes they use for cover, and outside of these shelters on mild days. They eat insects and arachnids, and occasionally lizard eggs. Mating occurs from late spring to early summer; females are pregnant with one or two young all summer, and they give birth to live young in the early fall.

Range and Variations: Granite Night Lizards live in western Riverside County, much of inland San Diego County, and small areas of western Imperial County. From there, they range into northern Baja California. Sandstone Night Lizards have one of the smallest ranges of any lizard, occupying only one tiny area within Anza-Borrego Desert State Park in San Diego County.

How to Find Sandstone and Granite Night Lizards: To find Granite Night Lizards, hike around in areas where flat pieces of granite have flaked off of large boulders to create the lizards' hiding spots underneath. On warm summer nights, use your flashlight to find the lizards crawling around on the boulders. During the day, your best bet is to carefully lift granite flakes to look for lizards hiding underneath. To find Sandstone Night Lizards, go to the Truckhaven Rocks area of Anza-Borrego Desert State Park and walk around at night looking for the lizards on the rocks. Note that you should not try to catch Sandstone Night Lizards because they are a species of special concern and because they occur only within a state park, where handling animals is prohibited. It is illegal to break apart rocks in search of any reptiles in California, so please leave these lizards' shelters exactly as you found them.

Chad Lane

ISLAND NIGHT LIZARD
XANTUSIA RIVERSIANA

FAMILY XANTUSIIDAE

Island Night Lizards are an example of a success story, where a previously highly threatened species recovered enough to be removed from the list of threatened and endangered species. They only live on a few islands off the shore of California, and their habitat was altered by the introduction of grazing livestock. On San Clemente Island, these lizards are associated with plants like California box thorn and various types of prickly pear, both of which were demolished by the goats that were introduced in the 1800s. The lizards also took a hit from feral cats on the island. However, the US Navy, which owns the island, removed goats and cats in the 1900s, and lizard populations have rebounded dramatically. And I mean *dramatically*. An in-depth study of the species on

San Clemente Island in the 1980s concluded that there can be as many as 3,200 lizards in an area about the size of a baseball field, which is a record for ground-dwelling lizards. Most of the other Channel Islands have similar histories where native flora and fauna were threatened by livestock, cats, and rodents introduced by people, then later recovered after they were removed. Island species like the Island Night Lizard are especially vulnerable because of their small ranges. Luckily, much of their habitat is now protected, by the navy on San Clemente and San Nicolas Islands and by the National Parks Service on Santa Barbara and Sutil Islands.

Appearance: Island Night Lizards are medium in size, with pointy snouts and lidless eyes with vertical pupils. Like other night lizards, they have granular skin on their bodies and large, smooth scales on their heads. Island Night Lizards have incredibly variable colorations and patterns, generally characterized by a mottled mixture of grays, whites, and browns. Sometimes the mottled spots can join to form stripes down their backs.

Natural History: Island Night Lizards are active during the day and occupy many habitats on the islands in which they live. They are secretive lizards that spend a lot of time in areas with hiding places, which can include dense vegetation, cover objects, and rock crevices. They eat spiders and insects, and they also eat a lot of plant matter. Mating takes place in the spring, and females give birth to live young in the early fall. Notably, Island Night Lizards are slow to attain reproductive maturity, and unlike most lizards, females only reproduce every two years.

Range and Variations: Island Night Lizards occur on a handful of California's Channel Islands: San Clemente Island, Santa Barbara

Island, Sutil Island, and San Nicolas Island. The lizards on San Nicolas Island are considered to be a separate subspecies from those on the other islands.

How to Find Island Night Lizards: Island Night Lizards are tricky to find, not because they are rare, but because access to the islands on which they live can be difficult to arrange. The only island of the three that is accessible to the public is Santa Barbara Island. You can take a boat trip from the coast to Santa Barbara Island from spring through fall, hike around for a few hours, then return to the mainland. Notably, Santa Barbara Island is part of the Channel Islands National Park, so wildlife cannot be captured, and rules about where you can hike may apply. If you go, I suggest asking a park ranger to help you choose a trail that runs through areas with California box thorn, prickly pear, and rocks. Keep your eyes out for lizards darting around under the vegetation or slinking into rock crevices. Binoculars are great for standing back and observing Island Night Lizards from the respectable distance that their protected status requires.

Chad Lane

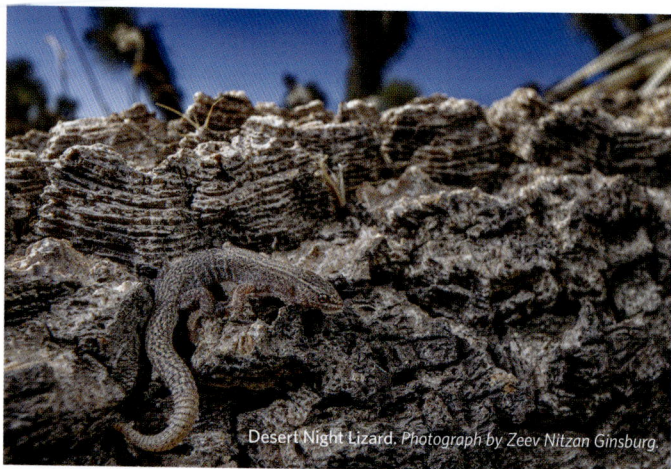
Desert Night Lizard. *Photograph by Zeev Nitzan Ginsburg.*

DESERT NIGHT LIZARD AND RELATIVES

XANTUSIA SIERRAE, X. VIGILIS, **AND** *X. WIGGINSI*

FAMILY XANTUSIIDAE

In the Mojave Desert, a tiny species of lizard called the Desert Night Lizard is the most common lizard you will never see . . . unless you know where to look. I recall being rather unimpressed by these lizards when I examined pickled specimens of the nondescript little things in my college herpetology lab. That all changed when we went to the Mojave Desert on a field trip and got to see these creatures in the wild. These "night" lizards are actually most active during the day, but this fact eluded scientists for years because the lizards' activity mostly takes place

Sierra Night Lizard.

Wiggins' Night Lizard.

out of sight, underneath and within fallen limbs of Joshua trees. Scientists were fooled by their vertical pupils, which were thought to indicate nocturnality (spoiler: they don't). It wasn't until a graduate student at UC Berkeley designed an ingenious apparatus to detect and record every time Desert Night Lizards moved about inside tiny cages, that their daytime habits were fully revealed. These lizards crawl around during the day under the bark and deep within the insides of rotting Joshua tree logs, hunting for the tiny insects that also shelter there. The logs are often lying in the shade of the living tree, and so this microhabitat represents a thermal oasis in the desert. The lizards cannot tolerate extreme heat and would rapidly perish if exposed to the outside. In my experience, I only ever see them sitting on the surface of the logs at dawn. Searching for any of these three species of lizards in the wild necessitates great know-how and care, as disturbing their shelters can mean permanently destroying their homes. The other two species used to be considered the same species as the Desert Night Lizard, and their habits are likely very similar even though they have not been studied in great detail.

Appearance: These three species look very similar. All are very small (their whole body and tail would fit in the palm of your hand) and narrow, with no moveable eyelids and with distinctive vertical pupils. They are gray or brown, with small dark spots that sometimes form a network of stripes down their backs. Their scales are small and granular, except on their head, where they are large and smooth. They resemble nocturnal geckos in terms of their size and eyes, but their pointy toes indicate that they do not climb about on smooth surfaces like geckos do with their splayed, sticky toes.

Natural History: As described above, night lizards are most active during the day, though you are unlikely to see them because their activity takes place mainly within fallen Joshua tree logs (or other cover objects) and associated rotting plant debris. All three species can occasionally be found in rock crevices, and the Sierra Night Lizard specializes in living in rocky outcrops. In spring and summer, they defend territories within these logs or outcrops, but in winter they congregate in family groups, where they have drastically reduced activity during cold months. Night lizards eat tiny insects and arachnids that are common under and within plant debris, including ants, termites, spiders, and others. Mating takes place in the summer, and females are pregnant usually with two young, which are born in the early fall.

Range and Variations: Desert Night Lizards (*Xantusia vigilis*) are found mainly in the Mojave Desert of California, from Mono County south to San Diego County, also extending west into San Luis Obispo and Santa Barbara Counties. They also occur in San Benito County, extending a bit into surrounding counties. Outside of California, Desert Night Lizards range in southern Nevada and Utah, western Arizona, and northern Mexico. Sierra Night Lizards (*Xantusia sierrae*) have a tiny range in the Sierra Nevada foothills in northern Kern County, where they occupy granite rock outcrops. The Wiggins' Night Lizard (*Xantusia wigginsi*) occurs mainly in Baja California but has populations in southeastern San Diego County.

How to Find Desert Night Lizards and Their Relatives: It is important to distinguish between how these lizards could be found and how they should be found. Because Desert Night Lizards typically live within fallen Joshua tree branches, scientists and lizard hunters of the past would often tear these branches apart to find the lizards within. However, this destroys their homes and is now considered to be a highly unethical practice. Instead, I recommend the following. Hike through Joshua tree forests during spring, summer, or early fall and look for living trees that cast shade over piles of dead branches below. Carefully lift the branches and watch for tiny lizards underneath. Be sure to replace the branches exactly as you found them, never tearing them apart. In many areas, these lizards are common enough that you will eventually find one if you stick with it. In winter they are likely to be deeper within the branches. Night lizards can also be found under other types of debris, and occasionally they hide in rock crevices, where you might find them by shining a flashlight inside on warm summer evenings. Notably, the Sierra Night Lizard is a rock specialist, living under granite flakes in its small range. Due to its protected status, you are prohibited from handling Sierra Night Lizards without a permit from the California Department of Fish and Wildlife. If you capture a Desert or Wiggins' Night Lizard, be very gentle with it and be sure not to expose it to hot conditions for long. It could even overheat from the warmth of your hand!

Northern Alligator Lizard. *Photograph by Spencer Riffle.*

ACKNOWLEDGMENTS

I am thankful to the legion of people who helped make this book a reality. First and foremost, one of the best aspects of this book is the beautiful photography. I am grateful to the photographers who allowed me to use their gorgeous images of California lizards: Patrick Alexander, Bryce Anderson, Brittany App, Sean Barefield, Chris DeGroof, Scott Eipper, William Flaxington, Zeev Nitzan Ginsburg, Lee Grismer, Francesca Heras, Adel Ibrahim, Marisa Ishimatsu, Brandon Kong, Chad Lane, Jeff Lemm, Grayson Lloyd, Jeff Martineau, Alex Mason, Nicolette Murphey, Jeff Nordland, Spencer Riffle, Max Roberts, Jackson Shedd, and Bill Walker. I owe a huge thanks to Marthine Satris, who asked only for one book on snakes and instead got another book forced on her . . . because the lizards wouldn't stand for the snakes having the last word. Emmerich Anklam helped me refine my language without stifling my joyful praise of lizards. The rest of the staff at Heyday are equally fabulous—thank you for packaging this book together into a product worthy of our state's scaly four-legged beasties. Big thanks to my first mentor, Harry Greene, who taught me how to tie a lizard lasso and took me on my first, life-changing field trip to the land of lizards that we called the Mojave Desert. I thank the people who generously read and critiqued all or parts of early versions of this book, including Robert Espinoza, Zeev Nitzan Ginsburg, Robert Hansen, Brandon Kong, Greg Pauly, and Jackson Shedd. I owe thanks for all the other things out there to my wonderful family, who have been supportive of every weird twist and turn I have followed on this trip called life. I especially thank my dashing and decidedly non-sauroblivious husband, Steve, who joins me in recognizing that life is better with lizards.

RECOMMENDED FURTHER READING

I hope that this book has awakened a love for lizards within you. There is always more to learn! Here are my recommendations for further reading to sate your curiosity about lizards.

SCIENTIFIC LITERATURE

I have curated a list of the scientific studies I read when writing this book. Find it at EmilyTaylorScience.com.

FIELD GUIDES

Hansen, Robert W. and Jackson D. Shedd, *California Amphibians and Reptiles* (Princeton Field Guides). Princeton University Press, 2025.

California finally has a dedicated field guide! This one is thorough, accurate, and is an absolute must-have for the serious field herpetologist.

Jones, Lawrence L. C. and Robert E. Lovich, eds., *Lizards of the American Southwest: A Photographic Field Guide*. Rio Nuevo Publishers, 2009.

For those who want a field guide only on lizards, this is your best bet. This book, with gorgeous photographs, focuses on lizards in the American Southwest, where biodiversity is high.

Stebbins, Robert C., *A Field Guide to Western Reptiles and Amphibians* (Peterson Field Guides). Houghton Mifflin Company, 2003.

Some lizard scientific names have changed since 2003, and California now has many more non-native lizards than they did then, but the beautiful paintings of Californian amphibians and reptiles make this field guide a collector's item.

www.CaliforniaHerps.com

I highly recommend this excellent, free, online field guide to the amphibians and reptiles of California, created by Gary Nafis.

BOOKS

There are many regional or specialty books on lizards. I have chosen two general books on lizards as recommendations for further reading. Enjoy!

Pianka, Eric R. and Laurie J. Vitt, *Lizards: Windows to the Evolution of Diversity*. University of California Press, 2003.

Rodda, Gordon H., *Lizards of the World: Natural History and Taxon Accounts*. Johns Hopkins University Press, 2020.

ABOUT THE AUTHOR

Born in a year of the dragon, Emily Taylor summons her most dragony energy as she shows off two young forest dragons captured on a recent expedition in a leech-infested jungle in Borneo. She is a Professor of Biological Sciences at the California Polytechnic State University in San Luis Obispo, California, where she conducts research on the physiology, ecology, and conservation biology of lizards and snakes with her students. She got her bachelor's degree in English at UC Berkeley and her PhD in Biology at Arizona State University. Her main Californian squeeze in terms of research is the Western Rattlesnake, but she also conducts numerous projects on its similarly ubiquitous legged brethren, the Western Fence Lizard. This is her second book; the first one was *California Snakes and How to Find Them*, also published by Heyday. She lives in Atascadero with her husband Steve in their madhouse of rescued creatures, including Pax the dog, Aperol Spritz the bearded dragon, Baby the Boa constrictor, Snakeholio and Fizz the rattlesnakes, and Flash, Helmut, and Bill the tortoises. Learn more at EmilyTaylorScience.com and follow her on social media @snakeymama.

Baja California Collared Lizard. *Photograph by Jackson Shedd.*